# Structure and Evolution
# of Single Stars

## An introduction

# Structure and Evolution of Single Stars

An introduction

**James MacDonald**

*University of Delaware*

Morgan & Claypool Publishers

Rights & Permissions
To obtain permission to re-use copyrighted material from Morgan & Claypool Publishers, please contact info@morganclaypool.com.

ISBN    978-1-6817-4105-5 (ebook)
ISBN    978-1-6817-4041-6 (print)
ISBN    978-1-6817-4233-5 (mobi)

DOI    10.1088/978-1-6817-4105-5

Version: 20151101

IOP Concise Physics
ISSN 2053-2571 (online)
ISSN 2054-7307 (print)

A Morgan & Claypool publication as part of IOP Concise Physics
Published by Morgan & Claypool Publishers, 40 Oak Drive, San Rafael, CA, 94903, USA

IOP Publishing, Temple Circus, Temple Way, Bristol BS1 6HG, UK

*To my stellar wife, Ruth, and children, Claire, Madolyn and Elena.*

# Contents

**Preface**                                                                    xiii

**Acknowledgements**                                                            xiv

**Author biography**                                                            xv

**1    Observational background**                                              **1-1**

1.1    Distances                                                               1-1
1.2    Stellar brightness and luminosity                                       1-2
1.3    Colors                                                                  1-3
1.4    Spectroscopy                                                            1-3
1.5    Color–magnitude diagrams                                                1-4
1.6    Stellar masses                                                          1-6
1.7    The mass–luminosity relation for main sequence stars                    1-8
1.8    The mass–radius relation for main sequence stars                        1-9
       Bibliography                                                            1-9

**2    The equations of stellar structure: mass conservation**                **2-1**
       **and hydrostatic equilibrium**

2.1    Introduction                                                            2-1
2.2    The mass conservation equation                                          2-1
2.3    The hydrostatic equilibrium equation for a spherical star               2-2
2.4    The dynamical time scale                                                2-3
2.5    The central temperature of the Sun                                      2-4
2.6    The central temperatures of main sequence stars                         2-5
2.7    Radiation pressure                                                      2-6

**3    Energy considerations, the source of the Sun's energy,**               **3-1**
       **and energy transport**

3.1    Introduction                                                            3-1
3.2    The virial theorem                                                      3-1
3.3    The virial theorem for stars in hydrostatic equilibrium                 3-3
3.4    The conservation of energy equation for a star in hydrostatic equilibrium   3-5
3.5    Stars in thermal equilibrium                                            3-6
3.6    Energy transport                                                        3-7
3.7    The equation of radiative transfer                                      3-8

3.8   Optical depth and effective temperature                          3-10
3.9   Validity of the diffusion approximation                          3-10
      Bibliography                                                      3-11

**4     Convective energy transport                                     4-1**

4.1   Introduction                                                      4-1
4.2   The Schwarzschild criterion for convective instability           4-1
4.3   Including convective energy transport in stellar models          4-4
      Bibliography                                                      4-6

**5     The equations of stellar evolution and how to solve them        5-1**

5.1   Introduction                                                      5-1
5.2   The equations of stellar structure                               5-1
5.3   The physical significance of the Eddington luminosity            5-3
5.4   Equations for composition changes                                5-3
5.5   Solving the equations of stellar evolution                       5-4
5.6   The Newton–Raphson method                                        5-5
5.7   Sets of non-linear equations                                     5-5
      Bibliography                                                      5-6

**6     Physics of gas and radiation                                    6-1**

6.1   Introduction                                                      6-1
6.2   The ideal gas equation of state                                  6-2
6.3   The radiation equation of state                                  6-3
6.4   The equation of state for a mixture of ideal gas and radiation   6-6
6.5   The Eddington standard model of stellar structure                6-7
      Bibliography                                                      6-8

**7     Ionization and recombination                                    7-1**

7.1   Introduction                                                      7-1
7.2   The Boltzmann excitation equation                                7-1
7.3   The Saha ionization equation                                     7-2
7.4   A difficulty and its resolution                                  7-3
7.5   Ionization of hydrogen                                           7-4
7.6   The effect of ionization on the adiabatic gradient               7-5
7.7   The effect of ionization on the specific heat                    7-7
7.8   Pressure ionization                                              7-7

7.9   Free energy approach to ionization                                        7-7

7.10  A crude model for inclusion of pressure ionization in a                   7-9
      thermodynamically consistent way

      Bibliography                                                              7-11


**8     The degenerate electron gas**                                           **8-1**

8.1   Introduction                                                              8-1

8.2   Complete electron degeneracy                                             8-1

8.3   Limiting forms                                                            8-4

8.4   The contribution from nuclei at zero temperature                         8-4

8.5   Transition from non-degeneracy to degeneracy                             8-5

8.6   Effects of degeneracy on the adiabatic gradient and the first            8-5
      adiabatic exponent


**9     Polytropes and the Chandrasekhar mass**                                 **9-1**

9.1   Introduction                                                              9-1

9.2   The Lane–Emden equation                                                  9-1

9.3   Application to white dwarf stars                                         9-3

      Bibliography                                                             9-4


**10    Opacity**                                                               **10-1**

10.1  Introduction                                                             10-1

10.2  The Rosseland mean opacity                                               10-1

10.3  Opacity mechanisms                                                       10-2

10.4  Electron scattering opacity                                              10-3

10.5  Free–free opacity                                                        10-3

10.6  Bound–free opacity                                                       10-4

10.7  Bound–bound opacity                                                      10-5

10.8  The Rosseland mean opacity for solar composition material               10-7

      Bibliography                                                             10-9


**11    Nuclear reactions**                                                     **11-1**

11.1  Introduction                                                             11-1

11.2  Occurrence of thermonuclear reactions                                    11-1

11.3  Cross sections and nuclear reaction rates                                11-2

11.4  The cross section                                                        11-4

11.5  Evaluation of the reaction rate                                          11-5

11.6 Major nuclear burning stages in stars: H burning 11-7

11.7 Energy generation in the pp-chains and the CNO-cycles 11-8

11.8 Major nuclear burning stages in stars: He burning 11-10

11.9 Advanced nuclear burning phases 11-11

Bibliography 11-12

**12 Neutrino energy loss processes** **12-1**

12.1 Pair annihilation neutrino process ($e^+ + e^- \rightarrow \nu + \bar{\nu}$) 12-1

12.2 Plasma neutrino process ($\gamma_{\text{plasmon}} \rightarrow \nu + \bar{\nu}$) 12-1

12.3 Photo-neutrino process ($\gamma + e \rightarrow e + \nu + \bar{\nu}$) 12-3

12.4 Bremsstrahlung neutrino process 12-3

Bibliography 12-4

**13 Homology relations** **13-1**

13.1 Introduction 13-1

13.2 Homology of zero age main sequence stars 13-1

13.3 Sensitivity of stellar structure to nuclear reaction rate 13-4

13.4 Sensitivity of stellar properties to composition 13-5

13.5 Stars with convective cores 13-6

13.6 Stars with convective envelopes 13-6

**14 Hydrogen main sequence stars** **14-1**

14.1 Masses of main sequence stars 14-1

14.2 Lifetimes of main sequence stars 14-1

14.3 Convection in main sequence stars 14-2

14.4 Variation of surface properties with mass 14-4

14.5 Variation of central properties with mass 14-5

14.6 The theoretical Hertzsprung–Russell diagram 14-8

Bibliography 14-8

**15 Helium main sequence stars** **15-1**

15.1 Why consider helium main sequence stars? 15-1

15.2 Homology analysis of helium zero age main sequence stars 15-1

15.3 Convection in helium main sequence stars 15-3

15.4 Variation of surface properties with mass 15-3

15.5 Variation of central properties with mass 15-4

15.6 The theoretical Hertzsprung–Russell diagram 15-7

Bibliography 15-7

## 16 The Hayashi line — 16-1

16.1 Introduction — 16-1
16.2 The Hayashi phase — 16-2
    Bibliography — 16-7

## 17 Star formation — 17-1

17.1 Introduction — 17-1
17.2 The Jeans mass — 17-1
17.3 Fragmentation — 17-5
    Bibliography — 17-5

## 18 Evolution on the main sequence and beyond — 18-1

18.1 Introduction — 18-1
18.2 Change in luminosity on the main sequence — 18-1
18.3 Evolution of the hydrogen profile — 18-3
18.4 Evolution after hydrogen exhaustion in the core — 18-3
18.5 The Hertzsprung gap — 18-5
    Bibliography — 18-11

## 19 Evolution on the red giant branch — 19-1

19.1 Introduction — 19-1
19.2 Change in luminosity on the red giant branch — 19-1
19.3 The globular cluster luminosity function bump — 19-3
19.4 The helium core flash — 19-5
19.5 Stability considerations — 19-6
    Bibliography — 19-8

## 20 Evolution from red giant to white dwarf — 20-1

20.1 Introduction — 20-1
20.2 The horizontal branch — 20-1
20.3 The asymptotic giant branch — 20-3
20.4 The formation of planetary nebulae — 20-4
20.5 The cooling of white dwarfs — 20-5
20.6 The luminosity function of white dwarfs — 20-8
20.7 Masses of white dwarf stars: observational material — 20-10
    Bibliography — 20-12

**21   Evolution of massive stars**                                          **21-1**

21.1  Introduction                                                             21-1
21.2  Composition changes in the core                                          21-5
21.3  Evolution after the end of core helium burning                           21-5
21.4  Evolution of stars more massive than 8 $M_\odot$                         21-6
      Bibliography                                                            21-11

# Preface

This book is an outgrowth of the notes for my course on stellar astrophysics which I have taught at the University of Delaware on a regular basis over the past 15 years. It is geared towards undergraduate students in their senior year and graduate students beginning their astronomical studies. I assume that students are comfortable with material from classical mechanics, quantum mechanics, statistical physics, and thermodynamics.

I make significant use of figures based on computations of stellar structure and evolution using my version of what is often referred to as the Eggleton code and I am grateful to my PhD adviser, Peter P Eggleton, for providing me with an early version of his easily adapted code. For students who wish to perform their own stellar evolution computations I recommend that they use the modern and efficient Modules for Experiments in Stellar Astrophysics (MESA) code written by Bill Paxton and colleagues. This code is currently downloadable from http://mesa.sourceforge.net/. A benefit of using this code is that it is well supported through a MESA-users mailing list and the MESA community portal.

# Acknowledgements

I wish to thank Dr Douglas Gough for introducing me to the fascinating topic of stellar evolution, Dr Peter Eggleton for providing me with his flexible stellar evolution code, and all my colleagues who over the years have added to my knowledge and experience, and in particular Drs Icko Iben, James Truran, and Sumner Starrfield.

# Author biography

**James MacDonald**

The author received his PhD in Astronomy from Cambridge University in 1979. Following postdoctoral positions at the Universities of Sussex and Illinois and Arizona State University, he joined the University of Delaware in 1985 where he is now a Professor of Physics and Astronomy. His scientific expertise is the study of the structure and evolution of stars. Recent work has focused on low mass main sequence stars and brown dwarfs. He has published over 80 papers in peer-reviewed journals.

# Chapter 1

## Observational background

### 1.1 Distances

One of the most difficult tasks in astronomy is finding accurate distances to objects. Distances to nearby stars can be found by trigonometric parallax. This method is shown schematically in figure 1.1.

As the Earth orbits the Sun, a nearby star will appear to move relative to the background of distant stars. This allows measurement of the parallax, $\theta$. The parallax of Proxima Centauri, the nearest star to the Sun, is 0.762 arc seconds. Since this, the largest parallax, is a very small angle, an accurate approximation in finding the distance to the star, $d$, is to replace $\tan \theta$ by $\theta$. The distance to the star, $d$, is then given by

$$d = \frac{1}{\theta},\tag{1.1}$$

where the unit of $d$ is the parsec (pc, parallax seconds) and $\theta$ is measured in arc seconds.

In terms of more familiar units, 1 pc = $3.086 \times 10^{13}$ km = 3.262 lyr. The distance to Proxima Centauri is then 1/0.762 pc = 1.31 pc = 4.28 lyr = $4.05 \times 10^{13}$ km.

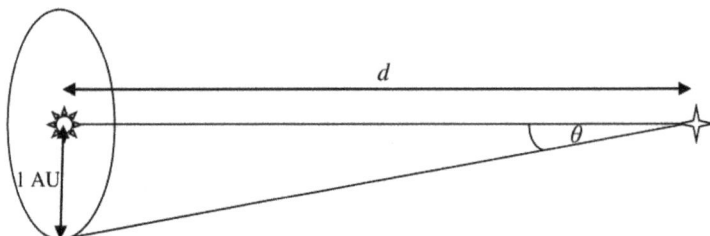

**Figure 1.1.** Schematic showing the definition of the parallax, $\theta$, for a star a distance, $d$, from the Sun.

doi:10.1088/978-1-6817-4105-5ch1

The trigonometric parallaxes of a large number of stars were measured with the Hipparcos satellite (Hipparcos is short for High Precision Parallax Collecting Satellite). The smallest measured parallax was about 1 milli-arc second, corresponding to a distance of 1 kiloparsec (1 kpc). The Gaia satellite, which was launched in December 2013, will measure the positions, distances and radial velocities of about one billion stars in our Galaxy and its satellite galaxies.

## 1.2 Stellar brightness and luminosity

Once we know the distance to a star we can convert the measured flux, i.e. the energy in some wavelength interval crossing unit area of a detector in unit time, to absolute luminosity, i.e. the energy emitted by the star in the same wavelength interval in unit time. (Alas, this is not so simple in practice because material along the line of sight to the star can scatter or absorb the starlight. A correction for *interstellar extinction* must then be made.)

The brightness of stars is usually measured on the magnitude scale (which originated with the ancient Greek astronomer Hipparchus). The apparent magnitude, $m$, of a star is related to the measured flux, $f$, in a particular wavelength interval (or band) by

$$m = -2.5 \log \frac{f}{f_0}, \tag{1.2}$$

where $f_0$ is a constant specific to the particular band. Clearly $f_0$ is equal to the flux of a star of zero magnitude. The absolute magnitude, $M$, of a star is defined to be the magnitude it would have if it were at a distance of 10 pc and there were no interstellar extinction. This is related to $m$ and $d$, the distance to the star in parsecs, by

$$M = m + 5 - 5 \log d - A, \tag{1.3}$$

where $A$ is the correction for interstellar extinction in magnitudes.

The bolometric magnitude is a measure of the total radiation from the star emitted over all wavelengths. The luminosity of a star is the total energy emitted (in electromagnetic radiation) in unit time. The luminosity, $L$, of a star and its absolute bolometric magnitude, $M_{\text{bol}}$, are related by

$$M_{\text{bol}} = 4.755 - 2.5 \log \frac{L}{L_\odot}, \tag{1.4}$$

where $L_\odot$ is the luminosity of the Sun. A recent measurement of the solar luminosity [1] is $3.827 \times 10^{33}$ erg s$^{-1}$.

The measurement of stellar radiation using filters is called multi-color photometry. There are a number of photometric systems in use. The 'standard' system [2, 3] makes use of three standard (transmission) filters called U, B, and V, which stand for ultra-violet, blue, and visual. The effective wavelengths of these filters are 365, 440, and 550 nm, respectively. The apparent magnitudes of a star measured with these filters are usually denoted by $U$, $B$ and $V$, rather than $m_{\text{U}}$, $m_{\text{B}}$ and $m_{\text{V}}$.

The difference between the bolometric magnitude and the visual magnitude is called the bolometric correction,

$$BC = M_{bol} - M_V. \tag{1.5}$$

Since only part of the total radiation is emitted in the visual part of the spectrum, the bolometric correction should be negative.

## 1.3 Colors

A difference between two magnitudes is called a color index or simply a color. The two colors in the UBV system are $U–B$ and $B–V$. A plot of one color against another for a set of stars is called a two color diagram. A plot of a color against magnitude for a set of stars is called a color–magnitude diagram (CMD). Plotting color against apparent magnitude is not very useful unless we have reason to believe that all the stars in the sample are at the same distance.

The physical significance of a color is that it is a measure of the temperature of the radiating surface. A cool piece of iron, e.g. at room temperature, emits radiation at infra-red wavelengths with peak emission at about 10 µm. If we heat the iron to about 1000 K, it will glow a dull red. If we continue to increase the temperature, it will first glow at a lighter red, then yellow, with peak emission shifting to shorter wavelengths. Although the surfaces of stars are usually mainly hydrogen gas or plasma, there is a similar qualitative relation between color and temperature. A cool star will emit more radiation in the V band than in the B band and hence $B–V$ will be positive. A hot star will emit more radiation in the B band than in the V band and hence $B–V$ will be negative. Because stars vary in other ways than surface temperature (e.g. surface gravity and composition), there is not a unique correspondence between a single color and temperature. Much of the degeneracy can be removed by considering two (or more) colors.

## 1.4 Spectroscopy

A lot more information can be obtained about a star by measuring its spectral energy distribution, i.e. how much energy is emitted at each wavelength. The composition of the surface material can (in principle) be determined from the spectral lines, together with the temperature (from the ratio of the strengths of lines of different excitation/ionization states of the same element) and surface gravity (from the width of the lines). However this is time consuming compared to photometry where many stars can be measured simultaneously using a CCD detector.

Two stars with very similar spectra are expected to have similar properties, including luminosity. This leads to the concept of spectroscopic parallax. Spectroscopic analysis shows that the surfaces of most stars are composed of mainly hydrogen and helium, with about ten hydrogen atoms (or ions) for every helium atom. Heavy elements contribute only a small part to the composition. However, this does not mean that the heavy elements are unimportant. They influence the stellar structure through their effects on opacity and nuclear reaction rates. (Heavy elements are those elements that were not produced in the big bang, and include carbon and heavier elements. All the heavy elements have been produced in stars.)

For stellar evolution studies, it is useful to specify the elemental abundances in terms of mass fractions. The mass fractions of hydrogen, helium, and the heavy elements are often denoted by $X$, $Y$, and $Z$, respectively. For example, in the outer layers of the Sun, 1 kg of the material is made up of about 0.732 kg of H, 0.253 kg of He, and 0.015 kg of heavy elements, so that the mass fractions are $X = 0.732$, $Y = 0.253$, and $Z = 0.015$ [4].

Stars with abundances similar to the Sun are called Population I stars (or Pop I stars). Stars with appreciably lower heavy-element abundances are called Pop II stars. There is also a kinematic difference between Pop I and Pop II stars. On average, Pop I stars have lower space velocities than Pop II stars. A related finding is that the scale height (the average distance of the stars from the Galactic plane) of Pop I stars in the solar neighborhood is less than that for Pop II stars. Globular cluster (GC) stars are Pop II stars. The GCs themselves are spherically distributed about the Galactic center.

Since the big bang did not produce any heavy elements, it is surmised that the first generation of stars had $Z = 0$. These are referred to as P III stars. No stars are known unequivocally to belong to Pop III. However a few stars of very low $Z$ have been discovered. For example, the star HE 0107-5240 [5] has $Z \sim 10^{-7}$. It is possible that this is a Pop III star, whose surface has been polluted by mass lost from other stars.

## 1.5 Color–magnitude diagrams

Since accurate distances to a large number of stars were obtained by the Hipparcos mission, it is instructive to look at the CMD for these stars, which is shown in figure 1.2.

The first feature that stands out is that the stars are not uniformly distributed in the CMD. Most stars lie in a diagonal band from bottom right to top left. This is called the main sequence (MS). We shall see later that this region is densely populated because a star spends most of its lifetime there.

A second striking feature is that there is a well-populated region branching off near the middle of the MS to the upper right. Because these stars are more luminous than MS stars of the same color (which is related to temperature), they must have a larger radiating area than the MS stars. Hence they are called 'giants'. Also, since they are cooler than MS stars of the same luminosity, they emit more of their radiation at long wavelengths and hence are 'redder' than the MS stars. They are therefore called 'red giants' and populate the red giant branch (RGB) of the CMD.

Note that there are a few stars that lie in a diagonal band below the MS. Because they are less luminous than MS stars of the same color they must have smaller radiating areas. Since they are hotter than MS stars of the same luminosity, they emit more of their radiation at short wavelengths and hence are 'bluer' than the MS stars. These stars are called white dwarf (WD) stars.

We can also consider the ranges of luminosity and temperature of the Hipparcos stars. The most luminous stars are about 10 000 times more powerful than the Sun. This does not mean that much more luminous stars do not exist, only that they are rare and there are not any within 1 kpc of the Sun. The faintest of the Hipparcos stars have a luminosity that is about 1% that of the Sun. However there are many

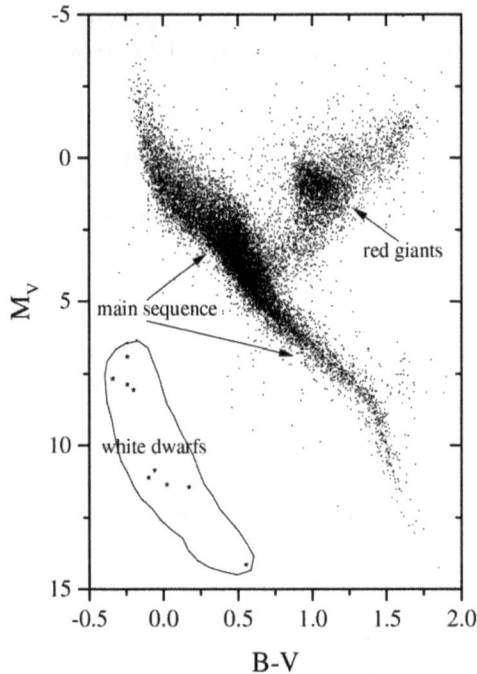

**Figure 1.2.** CMD for stars in the Hipparcos sample that have accurate parallaxes.

stars, possibly the majority, that are actually fainter than this. The reason why they do not appear in this diagram is simply that they are too faint for their parallax to be measurable by Hipparcos. This is an example of what is called a selection effect. The temperature range of the stars in the Hipparcos sample is from about 3000–30 000 K. Again there are many stars cooler than 3000 K, including the so-called brown dwarfs. The Gaia satellite is expected to remedy this situation and measure the properties of a few thousand brown dwarfs. In the Hipparcos sample, there are few MS stars with temperatures significantly higher than 30 000 K. However, as we shall see later, WD stars can be much hotter than this.

Notice that the MS is rather 'thick'. This is in part because the stars do not all have the same composition, particularly in the amount of heavy elements. Unresolved binary stars also contribute to the width of the MS[1].

In addition to the Hipparcos stars, it is instructive to plot a CMD for stars in a cluster. It is reasonable to assume that all the stars in the cluster are essentially at the same distance. Often it is also assumed that the stars all formed from a single gas cloud and hence have the same age and composition.

---

[1] Historical note: CMDs are also often referred to as Hertzsprung–Russell diagrams (HRDs). Hertzsprung and Russell independently plotted absolute magnitude against spectral type for stars near the Sun. Spectral type is also related to temperature and hence CMDs and HRDs contain similar information. Theorists call plots of luminosity against temperature HRDs.

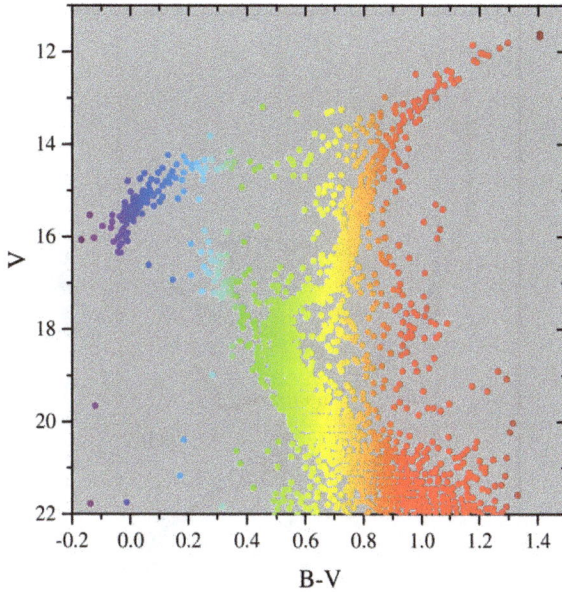

**Figure 1.3.** CMD for the GC M55. Data from [6, 7].

The CMD for the GC M55 is shown in figure 1.3. We can see more clearly how the RGB is related to the MS. The MS does not extend to the highest luminosities but has a 'turn off'. Except for a few stars known as 'blue stragglers', there are no MS stars more luminous than the turn off. We can conclude that any stars that were on the upper MS have evolved away from the MS. In addition, we see that there is a branch extending almost horizontally to the right from the giant branch. This horizontal branch (HB) is absent in the Hipparcos CMD but is seen in the CMDs of other GCs. (This difference is due to the lower heavy-element abundances of the GCs compared to the Sun. The stars equivalent to the GC HB form a clump near the RGB in the Hipparcos CMD.)

Figure 1.4 is a composite of cluster CMDs. There is a well-defined MS turn off (MSTO) for each cluster but they occur at different luminosities. As we will see later, this is a consequence of the clusters having different ages. Young clusters have MSTOs at high luminosities. As clusters age, the MSTO moves to lower luminosity. Also, the twin clusters h Persei and χ Persei contain large numbers of young stars that have not had time to reach the MS. This pre-MS (PMS) branch for these clusters connects to the MS near $B-V = 0$.

## 1.6 Stellar masses

Accurate stellar masses can be obtained for stars in some kinds of binary systems, namely visual binaries and double-line eclipsing binaries.

In visual binaries, the positions of the stars can be directly measured as they orbit their common center-of-mass. Since the center-of-mass lies at a focus of the elliptical orbit, the inclination of the orbit can be obtained. The radial component of the velocity

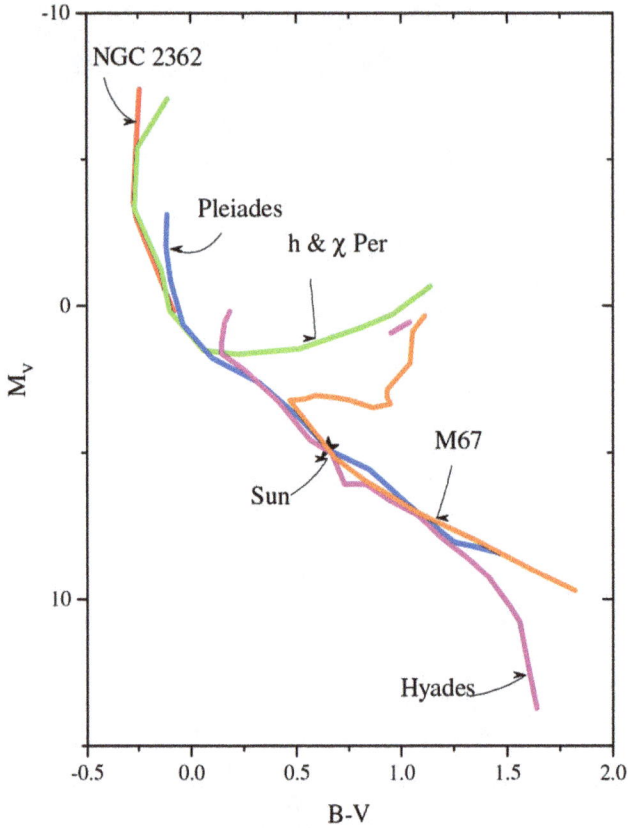

**Figure 1.4.** Composite CMD for a few open clusters. Data from [8] (NGC2362), [9] (h and χ Per), [10] (Pleiades), [11] (Hyades), and [12] (M67).

of each star (if bright enough) can be obtained by measuring the Doppler shift of the spectral lines. Using the inclination, these can be converted to orbital velocities. Combining the orbital velocities with the measured orbital period gives the absolute dimensions of the orbits (semi-major axes). Kepler's third law then gives the combined mass of the two stars. The mass ratio is the inverse of the ratio of the semi-major axes. Hence the mass of each component can be obtained. Furthermore, a comparison of the absolute dimensions with the apparent angular dimensions gives the distance to the binary. This allows apparent magnitudes to be converted to absolute magnitudes and hence luminosities.

In double-line eclipsing binaries, the inclination of the orbit is obtained from the eclipse light curve. The masses can then be obtained in the same way as for the visual binaries. If the distance is not known from trigonometric parallax, a distance estimate can be obtained from the eclipse durations, which give the sizes of the eclipsing stars. If the temperatures of the stars can be obtained from their spectra, combining with the stellar radii gives the stellar luminosity. The distance is then obtained by comparing with the apparent magnitudes.

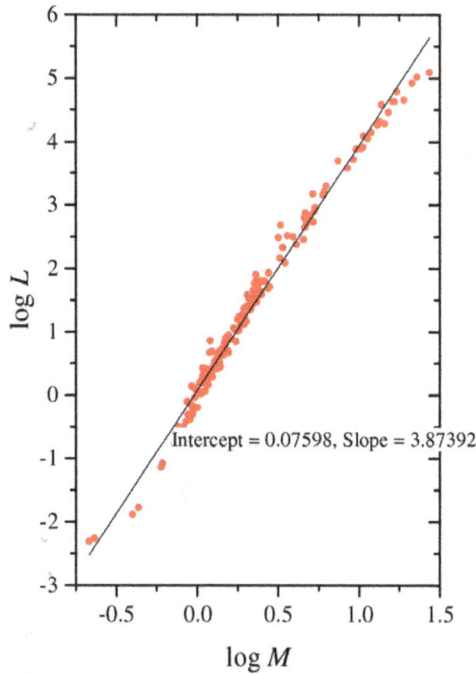

**Figure 1.5.** Mass–luminosity relation for stars with accurate masses. Data from [13].

## 1.7 The mass–luminosity relation for main sequence stars

As indicated above, the visual and double-line eclipsing binaries allow the luminosities of the stars to be found in addition to their masses. (The spectra also indicate the spectral type of the stars and hence we can determine whether they lie on or off the MS.) In figure 1.5, the logarithm of the luminosity in solar units is plotted against the logarithm of the mass in solar units for a sample of MS stars in visual or double-line eclipsing binaries. The slope of the regression line is approximately 4. Hence for MS stars (at least for those in the mass range 1/5–30 $M_\odot$), we find

$$L \propto M^4. \tag{1.6}$$

This is an important result. If we make the reasonable assumption that the energy radiated by the star is supplied by consumption of a 'fuel' which releases a fixed amount of energy per unit mass, then the lifetime of the fuel supply, $\tau$, is such that

$$\tau \propto \frac{M}{L} \propto M^{-3}. \tag{1.7}$$

Hence more massive stars have shorter lifetimes. This simple result goes a long way in explaining the composite CMD in figure 1.4. The cluster NGC 2362 is relatively young and has a MS populated by stars of a range of masses, including massive stars. The Pleiades cluster is older than NGC 2362 and its MS terminates at a lower luminosity because the most massive and hence most luminous stars have

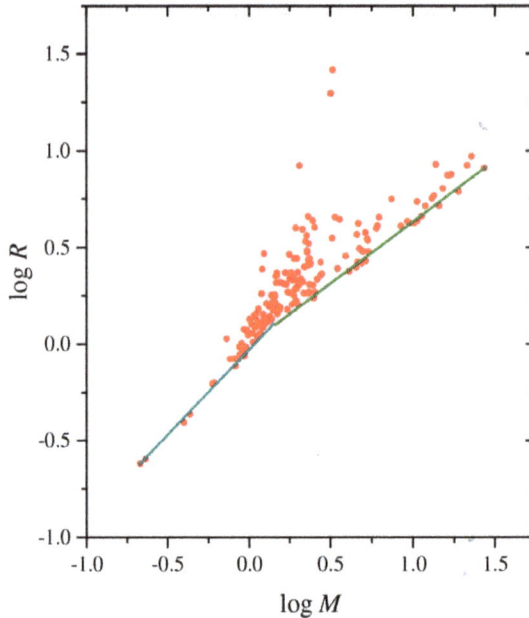

**Figure 1.6.** Mass–radius data for stars with accurate masses and radii.

consumed all their fuel and have 'died'. M67 is older still and stars only slightly more massive than the Sun have turned off the MS.

## 1.8 The mass–radius relation for main sequence stars

The double-line eclipsing binaries also provide accurate measurements of stellar radii. The mass–radius data for these stars are shown in figure 1.6. We see that for masses greater than 0.9 $M_\odot$, the spread in radius at a given mass is quite large. As we will see later, this is mainly due to the stars increasing in radius as they evolve on the MS and beyond (see figure 1.2). The straight lines represent how the radii of zero age MS (ZAMS) stars depend on their mass for lower mass stars (0.2 $M_\odot < M < 1.4\ M_\odot$) and higher mass stars (1.4 $M_\odot < M < 30\ M_\odot$). We see that the mass–radius relation for the lower mass stars, $M \propto R^{0.9}$, is steeper than for the higher mass stars, $M \propto R^{0.6}$, which suggests that there is a difference in the fundamental physics of low and high mass stars.

## Bibliography

[1] Kopp G and Lean J L 2011 *Geophys. Res. Lett.* **38** L01706
[2] Johnson H L and Morgan W W 1951 *Astrophys. J.* **114** 522
[3] Johnson H L and Morgan W W 1953 *Astrophys. J.* **117** 313
[4] Caffau E *et al* 2011 *Sol. Phys.* **268** 255–69
[5] Christlieb N *et al* 2004 *Astrophys. J.* **603** 708
[6] Vargas Álvarez C A and Sandquist E L 2007 *Astron. J.* **134** 825
[7] Kaluzny J *et al* 2010 *Acta Astron.* **60** 245

[8]  Perry Ch L 1973 *Spectral Classification and Multicolour Photometry* (*IAU Symp*. 50) p 192
[9]  Slesnick C L, Hillenbrand L A and Massey P 2002 *Astrophys. J.* **576** 880
[10] Johnson H L and Mitchell R I 1958 *Astrophys. J.* **128** 31
[11] Perryman M A C *et al* 1998 *Astron. Astrophys.* **331** 81
[12] Yadav R K S *et al* 2008 *Astron. Astrophys.* **484** 609
[13] Torres G, Andersen J and Giménez A 2010 *Astron. Astrophys. Rev.* **18** 67

# Chapter 2

## The equations of stellar structure: mass conservation and hydrostatic equilibrium

### 2.1 Introduction

To model stars, we make a number of basic assumptions. We assume that the star has spherical symmetry. This requires us to ignore the effects of rotation and magnetic fields. Except for exploding stars (such as supernovae) and pulsating stars, we can also assume that the star is in hydrostatic equilibrium (in which self-gravity is balanced by internal pressure). We also assume that Newtonian theory is an adequate description of gravity. This is reasonable because the escape velocity of most stars is much less than the speed of light in a vacuum. Einstein's general theory of relativity is needed for neutron stars and black holes.

The first two stellar structure equations that we will consider are those of mass conservation and hydrostatic balance.

### 2.2 The mass conservation equation

It is useful to consider two variables that give the location of a point inside the star. The radius variable, $r$, is the distance of the point from the stellar center and the mass variable, $m$, is the mass contained inside the sphere of radius $r$. By considering the mass in a thin spherical shell of inner radius $r$ and thickness $dr$, it is straightforward to show that

$$\frac{dm}{dr} = 4\pi r^2 \rho, \tag{2.1}$$

where $\rho$ is the mass density at radius $r$. This is the first of four stellar structure equations.

A useful related equation is

$$\frac{d\rho}{dt} + \rho \frac{1}{r^2} \frac{\partial(r^2 v)}{\partial r} = 0, \tag{2.2}$$

doi:10.1088/978-1-6817-4105-5ch2
2-1

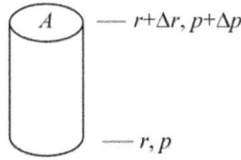

**Figure 2.1.** A small cylindrical volume element.

where d/d$t$ signifies a Lagrangian time derivative, i.e. the partial derivative with respect to time at a fixed value of $m$, and $v = $ d$r/$d$t$ is the velocity.

## 2.3 The hydrostatic equilibrium equation for a spherical star

The second stellar structure equation is the hydrostatic equilibrium equation. It is convenient to derive it by applying Newton's second law of motion to a small volume element. In particular, consider a small cylinder at radial position $r$, of cross sectional area $A$, and length $\Delta r$, with its axis in the radial direction, as shown in figure 2.1. By symmetry, the net force from the pressure on the curved surface of the cylinder is zero. The net force in the outward direction from the pressure at the ends of the cylinder is

$$pA - (p + \Delta p)A = -\Delta pA.$$

To find the radial component of the total force acting on the volume element, we need to include the downward-directed weight of the stellar material inside the cylinder. Hence the net force is

$$-\Delta pA - \rho \Delta rAg,$$

where $g$ is the acceleration due to gravity. (By convention, the vector acceleration $\mathbf{g} = -g\hat{\mathbf{r}}$, so that $g$ is positive.) This is equal to the mass of material inside the cylinder times its acceleration, so that

$$\rho \Delta rA \frac{\mathrm{d}^2 r}{\mathrm{d}t^2} = -\Delta pA - \rho \Delta rAg.$$

Dividing by $\Delta rA$ and letting $\Delta r \to 0$, we obtain

$$\rho \frac{\mathrm{d}^2 r}{\mathrm{d}t^2} = -\frac{\mathrm{d}p}{\mathrm{d}r} - \rho g. \tag{2.3}$$

In hydrostatic equilibrium, the acceleration term is negligible and is dropped from the equation. Furthermore, for a spherically symmetric object,

$$g = \frac{Gm}{r^2}. \tag{2.4}$$

Finally, we obtain the *equation of hydrostatic equilibrium*

$$\frac{\mathrm{d}p}{\mathrm{d}r} = -\rho \frac{Gm}{r^2}. \tag{2.5}$$

**Theorem.**
For a star of mass $M$ and radius $R$ in hydrostatic equilibrium, the central pressure, $p_c$, satisfies the inequality

$$p_c > \frac{GM^2}{8\pi R^4}.$$

**Proof.**
From the hydrostatic equilibrium and continuity equation, we have

$$-\frac{dp}{dm} = -\frac{dp/dr}{dm/dr} = -\frac{-Gm\rho/r^2}{4\pi r^2 \rho} = \frac{Gm}{4\pi r^4}.$$

Integrating over the star, we obtain

$$p_c - p_s = -\int_0^M \frac{dp}{dm}dm = \int_0^M \frac{Gm}{4\pi r^4}dm > \int_0^M \frac{Gm}{4\pi R^4}dm = \frac{GM^2}{8\pi R^4},$$

where $p_s$ is the pressure at the star's surface. Since $p_s > 0$, this proves the theorem.
    For the Sun, $M = 2 \times 10^{33}$ g, $R = 7 \times 10^{10}$ cm, and so $p_c > 5 \times 10^{14}$ dyne cm$^{-2}$ = $5 \times 10^8$ atmospheres. (Detailed models give $p_c = 2 \times 10^{17}$ dyne cm$^{-2}$.)

## 2.4 The dynamical time scale

To understand certain aspects of stellar evolution, it is useful to have an idea of various relevant time scales. One of these is the dynamical time scale. This can be estimated in a number of ways. One way is to consider a spherical star, of mass $M$ and radius $R$, in hydrostatic equilibrium (i.e. self-gravity is balanced by internal pressure) and suppose that the pressure instantaneously vanishes, so that the star collapses due to gravity. By applying conservation of mechanical energy to the motion of a point on the stellar surface, we have

$$\frac{1}{2}\left(\frac{dr}{dt}\right)^2 = \frac{GM}{r} - \frac{GM}{R}, \tag{2.6}$$

where $r$ is the radius of the star at time $t$. By making the substitution, $r = R\cos^2\phi$, this can be solved to obtain

$$\phi + \frac{1}{2}\sin 2\phi = \sqrt{\frac{2GM}{R^3}}\, t. \tag{2.7}$$

The time to reach zero radius, at $\phi = \pi/2$, is the dynamical time scale

$$t_{dyn} = \frac{\pi}{2}\sqrt{\frac{R^3}{2GM}}. \tag{2.8}$$

This can be written in terms of the mean density of the star, $\bar{\rho}$,

$$t_{dyn} = \sqrt{\frac{3\pi}{32G\bar{\rho}}}. \tag{2.9}$$

This is a useful result. It shows that if we can find dynamical information about a star, e.g. a fundamental mode pulsation period, then we can estimate the mean density of the star.

For the Sun, $\bar{\rho} = 1.4\,\mathrm{g\,cm^{-3}}$ and $t_{\mathrm{dyn}} \sim 30\,\mathrm{min}$. Since the Sun has not changed significantly over human history, we can deduce that the Sun is in overall hydrostatic equilibrium (to better than 1 part in $10^8$). (The outer parts of the Sun are in turbulent convective motion. Hence there are local deviations from strict hydrostatic equilibrium. However the spatial and temporal averages of the fluid acceleration are zero.)

An alternative way to estimate the dynamical time scale is to use the period of a test particle in an orbit that just grazes the stellar surface. This period is longer than the time scale given by equation (2.9) by a factor $\sqrt{32}$.

## 2.5 The central temperature of the Sun

We can use the equations of continuity and hydrostatic equilibrium to obtain a crude estimate of the temperature at the center of the Sun by assuming that the solar material obeys the ideal gas equation of state. The ideal gas law relates the pressure, $p$, to the particle number density, $n$, and temperature, $T$, by

$$p = nkT, \tag{2.10}$$

where $k$ is the Boltzmann constant. The numerical value is $k = 1.38 \times 10^{-16}\,\mathrm{erg\,K^{-1}}$.

The particle number density can be related to the mass density if we know the degree of ionization of each atomic species. Suppose species $i$ has, on the average, lost $Z_i$ electrons due to ionization. We have then that

$$n = \sum_i \frac{X_i\rho}{A_i m_{\mathrm{u}}}(1 + Z_i), \tag{2.11}$$

where $X_i$ and $A_i$ are the mass fraction and the atomic mass of species $i$. The ideal gas law can then be written as

$$p = \frac{k}{\mu m_{\mathrm{u}}}\rho T, \tag{2.12}$$

where the mean molecular weight (mass in amu per particle), $\mu$, is given by

$$\frac{1}{\mu} = \sum_i \frac{X_i}{A_i}(1 + Z_i). \tag{2.13}$$

The gas constant $\mathfrak{R} = k/m_{\mathrm{u}}$ has the numerical value $8.31 \times 10^7\,\mathrm{erg\,K^{-1}\,g^{-1}}$.

For fully ionized material of low heavy-element abundance,

$$\frac{1}{\mu} \approx \frac{5X + 3}{4}.$$

To estimate the central temperature of the Sun, assume that the density is uniform. The continuity equation gives

$$m = \frac{4\pi}{3}\rho r^3.$$

Using this in the hydrostatic equilibrium equation, we obtain

$$\frac{dp}{dr} = -\frac{4\pi}{3}G\rho^2 r,$$

so that

$$p = p_c - \frac{2\pi}{3}G\rho^2 r^2,$$

where $p_c$ is the central pressure.

Taking the pressure to be zero at the surface (where $r = R$), we have

$$p_c = \frac{2\pi}{3}G\rho^2 R^2 = \frac{3}{8\pi}\frac{GM^2}{R^4}.$$

Hence, assuming that the ideal gas law holds, the temperature at the center is

$$T_c = \frac{\mu}{\Re}\frac{p_c}{\rho_c} = \frac{\mu}{2\Re}G\frac{4\pi}{3}\rho R^2 = \frac{\mu}{2\Re}\frac{GM}{R}.$$

Assuming that for the Sun[1], $X_c = 0.35$ and that the material is fully ionized, we obtain $T_c \approx 10^7$ K. (Detailed solar modeling gives $T_c = 1.5 \times 10^7$ K.) Note that this temperature is much higher than the characteristic temperatures at which H and He become fully ionized ($10^4$ K and $10^5$ K, respectively) and hence the assumption of complete ionization is valid.

## 2.6 The central temperatures of main sequence stars

Assuming that the same physical assumptions hold for MS stars of different mass than the Sun, we can estimate their central temperatures from

$$T_c = 10^7 \frac{M}{M_\odot}\frac{R_\odot}{R} \text{ K}.$$

For the MS Hipparcos stars, we find empirically that

$$\frac{L}{L_\odot} \approx 14\left(\frac{T_e}{10^4 \text{ K}}\right)^6,$$

where $T_e$ is the effective temperature of the star (i.e. the equivalent blackbody temperature of the stellar surface). Using the scaling relations from the mass–luminosity relation and Stefan's law,

$$L \sim M^4,$$
$$L \sim R^2 T_e^4,$$

---

[1] It might seem more reasonable to take the central hydrogen mass fraction to be the same as at the surface. However, we now know that the Sun's energy comes from thermonuclear fusion of H to He and that the Sun is about halfway through consuming the H at its center.

we find that

$$R \sim M^{2/3}.$$

Hence

$$T_{\rm c} \approx 10^7 \left( \frac{M}{M_\odot} \right)^{1/3} {\rm K}.$$

We see that empirical relations indicate that more massive stars have higher central temperatures.

## 2.7 Radiation pressure

For isotropic radiation in thermal equilibrium with matter, the radiation pressure is

$$p_{\rm rad} = \frac{1}{3} a T^4, \qquad (2.14)$$

where the radiation constant, $a = 7.5657 \times 10^{-15}$ erg cm$^{-3}$ K$^{-4}$. Using the above relations for central temperature and central pressure, we find that

$$\frac{p_{\rm rad}}{p} = \frac{1}{3} a \frac{\left( \frac{\mu}{2\Re} \frac{GM}{R} \right)^4}{\frac{3}{8\pi} \frac{GM^2}{R^4}} = \frac{\pi}{18} a \left( \frac{\mu}{\Re} \right)^4 G^3 M^2 \approx 0.016 \left( \frac{M}{M_\odot} \right)^2,$$

at the stellar center.

Hence radiation pressure is more important for more massive stars. This crude estimate indicates that radiation pressure becomes dominant in MS stars of about 6 $M_\odot$. Detailed models indicate that radiation pressure cannot be neglected in MS stars more massive than about 10 $M_\odot$.

# Chapter 3

# Energy considerations, the source of the Sun's energy, and energy transport

## 3.1 Introduction

Here we consider some possible mechanisms for the source of the energy radiated from stars. For the Sun and other MS stars, we find that the energy comes from nuclear fusion reactions in the core of the star.

## 3.2 The virial theorem

The virial theorem provides a relationship between some global properties of a star. The virial is defined to be

$$I = \int_0^M r^2 \mathrm{d}m. \tag{3.1}$$

The virial theorem states that

$$\frac{1}{2}\frac{\mathrm{d}^2 I}{\mathrm{d}t^2} = 2K + \Omega + 3\int_0^M \frac{p}{\rho}\mathrm{d}m, \tag{3.2}$$

where $K$ is the kinetic energy of bulk motion of material in the star and $\Omega$ is the gravitational binding energy of the star.

One way to prove this theorem is by direct differentiation of the virial:

$$\frac{\mathrm{d}^2 I}{\mathrm{d}t^2} = \frac{\mathrm{d}^2}{\mathrm{d}t^2}\int_0^M r^2 \mathrm{d}m = \int_0^M \frac{\mathrm{d}^2}{\mathrm{d}t^2}(r^2)\mathrm{d}m = \int_0^M \left[2r\frac{\mathrm{d}^2 r}{\mathrm{d}t^2} + 2\left(\frac{\mathrm{d}r}{\mathrm{d}t}\right)^2\right]\mathrm{d}m. \tag{3.3}$$

The second term in the integral on the right is equal to $4K$. To evaluate the first term, we make use of the spherically symmetric conservation of momentum equation, derived in section 2.3,

doi:10.1088/978-1-6817-4105-5ch3          3-1

$$\rho\frac{\mathrm{d}^2 r}{\mathrm{d}t^2} = -\frac{\mathrm{d}p}{\mathrm{d}r} - \rho\frac{Gm}{r^2}.$$ (3.4)

This gives

$$\int_0^M r\frac{\mathrm{d}^2 r}{\mathrm{d}t^2}\mathrm{d}m = \int_0^M \frac{r}{\rho}\left(-\frac{\mathrm{d}p}{\mathrm{d}r} - \rho\frac{Gm}{r^2}\right)\mathrm{d}m = -\int_0^M \frac{r}{\rho}\frac{\mathrm{d}p}{\mathrm{d}r}\mathrm{d}m - \int_0^M \frac{Gm}{r}\mathrm{d}m.$$ (3.5)

The first integral on the right is

$$-\int_0^M \frac{r}{\rho}\frac{\mathrm{d}p}{\mathrm{d}r}\mathrm{d}m = -\int_0^M \frac{r}{\rho}\frac{\mathrm{d}p}{\mathrm{d}r}4\pi r^2\rho\,\mathrm{d}r = -\int_0^R \frac{\mathrm{d}p}{\mathrm{d}r}4\pi r^3\mathrm{d}r = \left[-p4\pi r^3\right]_0^R + \int_0^R 12\pi r^2 p\,\mathrm{d}r.$$

(3.6)

If we assume that the pressure at the surface is zero, the first term on the right is identically zero. Then

$$-\int_0^M \frac{r}{\rho}\frac{\mathrm{d}p}{\mathrm{d}r}\mathrm{d}m = \int_0^R 12\pi r^2 p\,\mathrm{d}r$$

$$= \int_0^M 12\pi r^2 p\frac{\mathrm{d}r}{\mathrm{d}m}\mathrm{d}m = \int_0^M 12\pi r^2 p\frac{1}{4\pi r^2\rho}\mathrm{d}m = 3\int_0^M \frac{p}{\rho}\mathrm{d}m.$$ (3.7)

The final term to consider is

$$\Omega = -\int_0^M \frac{Gm}{r}\mathrm{d}m.$$ (3.8)

This is the gravitational binding energy of the star. To see why, consider an isotropic sphere of mass $m$ and radius $r$. Outside the sphere the gravitational acceleration is

$$\mathbf{g} = -\frac{Gm}{s^2}\hat{\mathbf{n}},$$ (3.9)

where $s$ is the distance from the center of the sphere and $\hat{\mathbf{n}}$ is the outward normal. Hence outside the sphere the gravitational potential is

$$\Phi(s) = -\frac{Gm}{s},$$ (3.10)

where, by convention, the zero point for the potential is taken to be at infinity. (Note that this expression does not hold inside the sphere.) The gravitational potential at the surface of the sphere is

$$\Phi(r) = -\frac{Gm}{r}.$$ (3.11)

If we add a mass $\Delta m$ to the surface, the gravitational binding energy is incremented by

$$\Phi(r)\Delta m = -\frac{Gm}{r}\Delta m.$$ (3.12)

3-2

Hence by building up the star one shell at a time, we see that $\Omega$ is the gravitational binding energy of the star.

Putting the pieces together we have

$$\frac{1}{2}\frac{d^2I}{dt^2} = \int_0^M r\frac{d^2r}{dt^2}dm + 2K$$

$$= -\int_0^M \frac{r}{\rho}\frac{dP}{dr}dm - \int_0^M \frac{Gm}{r}dm + 2K = 3\int_0^M \frac{P}{\rho}dm + \Omega + 2K, \quad (3.13)$$

which is the virial theorem.

## 3.3 The virial theorem for stars in hydrostatic equilibrium

For stars in hydrostatic equilibrium, the kinetic energy is zero and also $I$ does not change with time. Hence

$$3\int_0^M \frac{P}{\rho}dm + \Omega = 0. \qquad (3.14)$$

For an ideal gas, the internal energy per unit mass is

$$u = \frac{3}{2}\frac{P}{\rho}. \qquad (3.15)$$

Hence the total internal energy of a star supported by ideal gas pressure is

$$U = \int_0^M u\,dm = \frac{3}{2}\int_0^M \frac{P}{\rho}dm. \qquad (3.16)$$

Using the virial theorem, we have for a star in hydrostatic equilibrium supported by ideal gas pressure

$$2U + \Omega = 0. \qquad (3.17)$$

The total energy of the star is

$$E = U + \Omega. \qquad (3.18)$$

Hence using equation (3.17),

$$E = -U = \frac{\Omega}{2}. \qquad (3.19)$$

Note that since $U$ is positive, the total energy is negative. This simply means that the star is bound.

The simple result in equation (2.3) has some remarkable implications. The total energy of a star can change if the energy lost by radiation at the surface is not balanced by internal sources of energy. Suppose that there are no energy sources. The rate of change of total energy is

$$\frac{dE}{dt} = -L, \qquad (3.20)$$

where $L$ is the luminosity of the star. Since $E = -U$, this means that as the star loses energy, its internal energy increases, i.e. it must become hotter! In other words, a star supported by thermal pressure has a negative specific heat. From $E = \Omega/2$, we also see that as the star loses energy, its binding energy becomes more negative, i.e. the star must have an overall contraction. We can associate a time scale with this heating/contraction by making the same crude approximation that was made to estimate the central temperature, i.e. the star has uniform density. In this case

$$\Omega = -\int_0^M \frac{Gm}{r}\mathrm{d}m = -\int_0^R \frac{G}{r}\frac{4\pi r^3\rho}{3}4\pi r^2\rho\mathrm{d}r = -\frac{16\pi^2}{3}G\rho^2\int_0^R r^4\mathrm{d}r$$

$$= -\frac{16\pi^2}{15}G\rho^2 R^5 = -\frac{3}{5}\frac{GM^2}{R}. \tag{3.21}$$

Hence

$$\frac{\mathrm{d}E}{\mathrm{d}t} = \frac{1}{2}\frac{\mathrm{d}\Omega}{\mathrm{d}t} = \frac{3}{10}\frac{GM^2}{R^2}\frac{\mathrm{d}R}{\mathrm{d}t} = -L. \tag{3.22}$$

The radius changes on a time scale

$$t_{\mathrm{th}} \approx \left|\frac{1}{R}\frac{\mathrm{d}R}{\mathrm{d}t}\right|^{-1} \approx \frac{GM^2}{RL}. \tag{3.23}$$

Because, from the virial theorem, this is also the time scale on which the internal energy of the star changes, it is called the *thermal time scale*. It is also often called the Kelvin–Helmholtz time scale, after the physicists who first considered whether the Sun could shine by releasing gravitational energy through contraction.

For the Sun,

$$t_{\mathrm{th}} \approx 3 \times 10^7 \mathrm{yr}.$$

Although this is a long time, 19th century geologists argued that it is much too short for the observed weathering of rocks and geological features to have occurred. Later radiological dating of Earth rocks and meteorites showed that the solar system and hence, by inference, the Sun were at least 4.5 Gyr old.

Hence we are forced to conclude that the Sun has an internal energy source. If this energy source was chemical in nature (i.e. the energy is stored in bonds between atoms), the Sun's MS lifetime would be about 7000 years. Because nuclear reactions release about $10^7$ times as much energy per unit mass as chemical reactions, we see that there is a potentially plentiful supply of nuclear fuel in the Sun.

In the conversion of H to He by nuclear processes about 0.007 of the rest mass energy is converted into heat and light. The nuclear lifetime of a star converting H to He (astronomer's loosely use the term hydrogen burning for this process) is

$$t_{\mathrm{nuc}} = \frac{0.007XMc^2}{L} = 10^{11}X\frac{M}{M_\odot}\left(\frac{L}{L_\odot}\right)^{-1} \mathrm{years}.$$

(We shall see later that the Sun and other stars consume only about 10% of their hydrogen while on the MS. Hence MS lifetimes are about a factor 10 smaller than the nuclear time scale.)

## 3.4 The conservation of energy equation for a star in hydrostatic equilibrium

Consider a spherical shell sandwiched between mass coordinates $m$ and $m + \Delta m$. The internal energy in the shell is $u\Delta m$. This can change because (i) radiation flows into and out of the shell, (ii) nuclear reactions can add energy to the shell, (iii) the pressure force can do work on the shell, and (iv) the gravitational force can do work on the shell. We will consider these processes in turn.

i.  Let the total outward flow of energy be $L$ at the inner boundary and $L + \Delta L$ at the outer boundary of the shell. (Here $L$ is a local quantity and not the luminosity of the star.) The change in internal energy in the shell in time $\delta t$ due to radiation is

$$\delta(u\Delta m)_1 = -\Delta L \delta t.$$

ii. Let nuclear reactions produce energy per unit mass at rate $\varepsilon$. The change in internal energy in the shell in time $\delta t$ due to nuclear energy production is

$$\delta(u\Delta m)_2 = \varepsilon \Delta m \delta t.$$

iii. The pressure force does work at both the inner and outer boundaries, which have radii $r$ and $r + \Delta r$. At the inner boundary, the total force is $4\pi r^2 p$, and if this boundary moves a distance $\delta r$ in time $\delta t$, the work done by this outward-directed force on the shell is $4\pi r^2 p \delta r = 4\pi r^2 p v \delta t$, where $v$ is the material velocity at radius $r$. Similarly, the work done by the inward-directed force at the outer boundary is $-4\pi(r + \Delta r)^2(p + \Delta p)(v + \Delta v)\delta t$. The change in internal energy due to the work done by the pressure force is

$$\delta(u\Delta m)_3 = -4\pi(r + \Delta r)^2(p + \Delta p)(v + \Delta v)\delta t + 4\pi r^2 p v \delta t = -\Delta(4\pi r^2 p v)\delta t.$$

iv. The change in internal energy due to the work done by gravity is

$$\delta(u\Delta m)_4 = -\frac{Gm\Delta m}{r^2}v\delta t.$$

Adding the four contributions, dividing by $\delta t$ and $\Delta m$, and taking the limits $\delta t \to 0$, and $\Delta m \to 0$, we obtain

$$\frac{du}{dt} = -\frac{dL}{dm} + \varepsilon - \frac{d(4\pi r^2 p v)}{dm} - \frac{Gm}{r^2}v = -\frac{dL}{dm} + \varepsilon - p\frac{d(4\pi r^2 v)}{dm} - 4\pi r^2 v\frac{dp}{dm} - \frac{Gm}{r^2}v$$

$$= -\frac{dL}{dm} + \varepsilon - \frac{p}{4\pi r^2 \rho}\frac{d(4\pi r^2 v)}{dr} - \frac{v}{\rho}\left(\frac{dp}{dr} + \rho\frac{Gm}{r^2}\right).$$

For a star in hydrostatic equilibrium, the last term is identically zero. The energy equation is then

$$\frac{du}{dt} = -\frac{dL}{dm} + \varepsilon - \frac{p}{\rho}\frac{1}{r^2}\frac{d(r^2 v)}{dr}. \tag{3.24}$$

The velocity can be eliminated by using the spherically symmetric form of the continuity equation, given in section 2.2,

$$\frac{d\rho}{dt} + \rho\frac{1}{r^2}\frac{d(r^2 v)}{dr} = 0,$$

to obtain

$$\frac{du}{dt} = -\frac{dL}{dm} + \varepsilon + \frac{p}{\rho^2}\frac{d\rho}{dt}.$$

This is more usually written as

$$\frac{dL}{dm} = \varepsilon - \frac{du}{dt} + \frac{p}{\rho^2}\frac{d\rho}{dt}, \tag{3.25}$$

which is the third equation of stellar structure.

The last two terms on the right-hand side of equation (3.25) are often grouped together and are called the gravo-thermal energy generation rate

$$\varepsilon_g = -\frac{du}{dt} + \frac{p}{\rho^2}\frac{d\rho}{dt}. \tag{3.26}$$

If the composition of the star is not changing, then the laws of thermodynamics allow us to express the gravo-thermal energy generation rate in terms of the rate of change of entropy

$$\varepsilon_g = -T\frac{ds}{dt}, \tag{3.27}$$

where $s$ is the entropy per unit mass. This form is useful for conceptual understanding of some aspects of stellar structure but is not of practical use, because the composition does change due to nuclear transformations and also because of turbulent mixing processes.

## 3.5 Stars in thermal equilibrium

A star is in thermal equilibrium if the gravo-thermal energy generation rate is zero everywhere. As a consequence, the radiative losses at the surface are balanced by nuclear energy sources in the interior. In thermal equilibrium

$$\frac{dL}{dm} = \varepsilon. \tag{3.28}$$

The luminosity at the surface is then

$$L_* = \int_0^M \frac{dL}{dm}dm = \int_0^M \varepsilon \, dm = \bar{\varepsilon}M,$$

where $\bar{\varepsilon}$ is the mean nuclear energy generation rate. For the Sun, $\bar{\varepsilon} \simeq 2 \text{ erg g}^{-1}\text{s}^{-1}$.

Since the mass–luminosity relation for MS stars is $L_* \sim M^4$, we see that $\bar{\varepsilon} \sim M^3$. Hence the mean nuclear energy generation rate is higher in more massive stars. We saw earlier that the central temperature of MS stars also increases with mass. An energy generation rate that increases with temperature is a characteristic of thermonuclear reactions and suggests that the energy radiated by MS stars comes from nuclear fusion. For nuclear fission, the rate is independent of temperature and hence the mean nuclear energy generation rate would be independent of mass.

## 3.6 Energy transport

There are three main ways in which energy moves in a star. These are the familiar processes of radiation, conduction and convection. (These are not the only mechanisms but are the most common in stellar interiors. Waves can also transport energy, and wave mechanisms are important for heating stellar chromospheres and coronae, and for driving cool star winds.) Energy transport by radiation and conduction are similar in that they depend on collisions of energetic particles with less energetic particles. They differ in the nature of the particle that carries the energy. Electrons are the dominant carrier for conduction, whereas photons are responsible for radiative transport.

In most stars, the gas pressure is greater than radiation pressure. The same is true for the internal energy densities. The internal energy per unit volume of a perfect gas is

$$\rho u_{\text{gas}} = \frac{3}{2}nkT, \tag{3.29}$$

and the internal energy per unit volume of radiation is

$$\rho u_{\text{rad}} = aT^4. \tag{3.30}$$

Hence

$$\frac{u_{\text{gas}}}{u_{\text{rad}}} = \frac{3}{2}\frac{nkT}{aT^4} = \frac{p_{\text{gas}}}{2p_{\text{rad}}}. \tag{3.31}$$

At the Sun's center, $p_{\text{gas}} = 2 \times 10^{17}$ dyne cm$^{-2}$ and $p_{\text{rad}} = 10^{14}$ dyne cm$^{-2}$. Thus, since the gas has a higher energy density than radiation, it might be expected that conduction is more important than radiation for energy transport. However there is another factor that influences the efficiency of energy transport. This is the *mean free path*, which is the average distance a particle travels between collisions. A larger mean free path gives a higher rate of energy transport.

The mean free path, $\lambda$, is related to the collision cross section, $\sigma$, by

$$\lambda n \sigma = 1, \tag{3.32}$$

where $n$ is the particle number density. (Since a particle travels a distance $\lambda$ before colliding with a particle of cross section $\sigma$ there is one particle in volume $\lambda\sigma$. There are $n$ particles in unit volume. Hence $\lambda n\sigma = 1$.)

For thermal particles with Coulomb interactions, we can estimate the cross section by finding the inter-particle distance at which the Coulomb force affects the particle trajectories. Let the charges on the two interacting particles be $Z_1 e$ and $Z_2 e$ (in electrostatic units). Since the typical kinetic energy of the particles is $\sim kT$, the Coulomb energy is comparable when the particle separation is

$$s \approx \frac{Z_1 Z_2 e^2}{kT}. \tag{3.33}$$

The cross section is then

$$\sigma \approx \pi s^2 \approx \pi \left( \frac{Z_1 Z_2 e^2}{kT} \right)^2 = 10^{-5} \left( \frac{Z_1 Z_2}{T} \right)^2 \text{ cm}^2, \tag{3.34}$$

where $T$ is in K. Hence, for a pure H plasma, the mean free path is

$$\lambda = \frac{1}{n\sigma} \approx 10^{-19} \frac{T^2}{\rho} \text{ cm}. \tag{3.35}$$

At the center of the Sun, $T \sim 10^7$ K and $\rho \sim 100$ g cm$^{-2}$, and so $\lambda \sim 10^{-7}$ cm.

The mean free path of a photon depends on the opacity of the material, which we will consider in more detail later. At the center of the Sun, most of the electrons are free. A major source of opacity comes from photons scattering off free electrons. The relevant cross section is the Thomson cross section[1]

$$\sigma_e = \frac{8\pi}{3} \left( \frac{e^2}{m_e c^2} \right)^2 = 6.7\,10^{-25} \text{ cm}^2. \tag{3.36}$$

This is much smaller than a typical cross section from Coulomb interactions and hence the mean free path for photons is much larger, $\lambda \sim 10^{-2}$ cm. As a consequence, in MS stars radiation is much more efficient than conduction at transporting energy.

## 3.7 The equation of radiative transfer

We can derive an approximation to the equation of radiative transfer by considering a simple picture of how radiative transfer works. Consider two planar, semi-infinite blackbodies, with a small temperature difference, separated by one mean free path, so that the photons can travel between the blackbodies before being scattered or absorbed (see figure 3.1).

---

[1] This is related to the classical electron radius, which is obtained from a classical model of the electron in which the mass of the electron arises solely from its electrostatic energy.

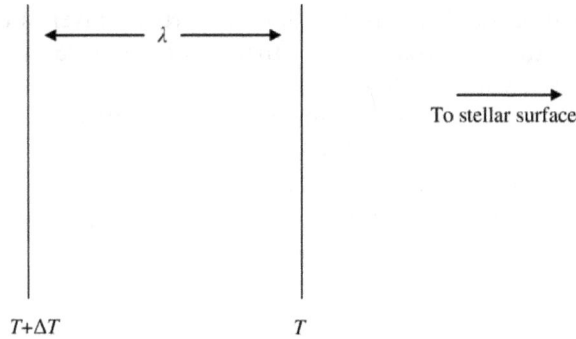

**Figure 3.1.** Schematic of radiative transfer between two surfaces separated by one photon mean free path.

The emission per unit area from a blackbody of temperature $T$ is $\sigma_B T^4$, where $\sigma_B = ac/4$ is the Stefan–Boltzmann constant. Since a blackbody absorbs all radiation falling on it, the heat transfer per unit area from the hotter to the cooler body is

$$H \approx \sigma_B(T + \Delta T)^4 - \sigma_B T^4 \simeq 4\sigma_B T^3 \Delta T. \tag{3.37}$$

Because

$$\Delta T = -\lambda \frac{dT}{dr}, \tag{3.38}$$

we have

$$H \approx -4\sigma_B T^3 \lambda \frac{dT}{dr}. \tag{3.39}$$

(The minus sign arises because heat flows from the hotter to the cooler body, i.e. down the temperature gradient.)

The *opacity*, $\kappa$, is defined by

$$\kappa\rho = n\sigma = \frac{1}{\lambda}. \tag{3.40}$$

Hence

$$H \approx -\frac{4\sigma_B T^3}{\kappa\rho}\frac{dT}{dr}. \tag{3.41}$$

A more detailed calculation (see e.g. [1]) shows that this is incorrect by a factor of 4/3. The correct result is that the radiative heat flux is

$$H = -\frac{4acT^3}{3\kappa\rho}\frac{dT}{dr}, \tag{3.42}$$

where we have used the result that $\sigma_B = ac/4$.

The radiative luminosity is the total energy carried by radiation through the surface of a sphere of radius $r$,

$$L_{\text{rad}} = 4\pi r^2 H = -\frac{16ac\pi r^2 T^3}{3\kappa\rho}\frac{\mathrm{d}T}{\mathrm{d}r}. \tag{3.43}$$

In deriving this result, we have used the *diffusion approximation* by making the assumption that the mean free path is much less than the length scale over which the temperature changes.

## 3.8 Optical depth and effective temperature

The optical depth, $\tau$, is a dimensionless measure of integrated absorptivity along the line of sight. It is defined by

$$\frac{\mathrm{d}\tau}{\mathrm{d}r} = -\kappa\rho. \tag{3.44}$$

The minus sign is so that the optical depth increases inwards. Since $\kappa\rho\lambda = 1$, we see that the surface of last 'scattering' occurs near optical depth near unity. A detailed treatment of radiative transfer shows that (for plane parallel atmospheres) the temperature of the stellar material is equal to that of an equivalent blackbody at optical depth 2/3. This is less than 1 because photons do not all leave the star in the perfectly radial direction.

The temperature of the equivalent blackbody is called the *effective temperature*. Provided the atmosphere is thin, the effective temperature, $T_{\text{eff}}$, is given by

$$L_* = 4\pi R^2 \sigma_B T_{\text{eff}}^4 = \pi ac R^2 T_{\text{eff}}^4. \tag{3.45}$$

The region of the star from which the observable photons are emitted is called the *photosphere*. Hence $T_{\text{eff}}$ is a measure of the temperature of the photosphere.

## 3.9 Validity of the diffusion approximation

The temperature scale height is

$$H_T = \left(-\frac{\mathrm{d}\ln T}{\mathrm{d}r}\right)^{-1} = \frac{16ac\pi r^2 T^4}{3\kappa\rho L_{\text{rad}}}. \tag{3.46}$$

Hence the ratio of photon mean free path to temperature scale height is

$$\frac{\lambda}{H_T} = \frac{1}{\kappa\rho H_T} = \frac{3L_{\text{rad}}}{16ac\pi r^2 T^4}. \tag{3.47}$$

Using equation (3.45), we find at the photosphere,

$$\frac{\lambda}{H_T} = \frac{3}{16},$$

and so we see that the diffusion approximation is reasonable even in the low density photospheric regions of the star. The diffusion approximation becomes better at higher optical depth.

## Bibliography

[1] Mihalas D 1978 *Stellar Atmospheres* 2nd edn (San Francisco, CA: W H Freeman)

Structure and Evolution of Single Stars
An introduction
**James MacDonald**

# Chapter 4

## Convective energy transport

### 4.1 Introduction

Convection is due to an instability that arises when the temperature gradient exceeds a critical value. There is as yet no simple but accurate way of including convective energy transport in models of stars. A phenomenological model of convection called *mixing length theory* is most often used.

### 4.2 The Schwarzschild criterion for convective instability

To derive the conditions under which convective instability will occur, we consider a region of the star that is in radiative equilibrium. Suppose a small volume element, shown by a circle in figure 4.1, at height $z$ is moved to height $z + \Delta z$ slowly enough that it remains in pressure balance with its surroundings, but quickly enough that it does not lose energy to its surroundings. This means that the entropy of the element does not change, i.e. it is moved adiabatically.

Because the element remains in pressure balance with its surroundings, the changes in pressure in the element and the surroundings are both given by

$$\delta p = \Delta p = \frac{dp}{dz}\Delta z. \tag{4.1}$$

Because the element is moved adiabatically, the change in the pressure inside the element and its density are related by

$$\frac{\delta p}{p} = \Gamma_1 \frac{\delta \rho}{\rho}, \tag{4.2}$$

where $\Gamma_1$ is the first adiabatic exponent (this expression is actually the definition of the first adiabatic exponent). $\Gamma_1$ is, in principle, a known function of density, temperature, and composition.

doi:10.1088/978-1-6817-4105-5ch4     4-1

Element                                    Surroundings

—————————◯———————————————————— $z+\Delta z$

$p+\delta p,\ \rho+\delta\rho$                                 $p+\Delta p,\ \rho+\Delta\rho$

—————————◯———————————————————— $z$

$p,\ \rho$                                        $p,\ \rho$

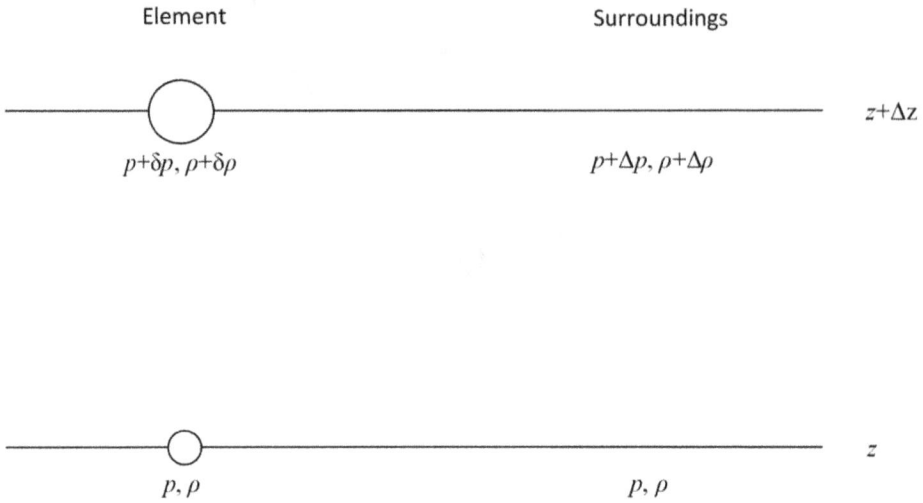

**Figure 4.1.** Schematic used in deriving the criterion for convective instability.

The change in density of the element is then

$$\delta\rho = \frac{\rho}{\Gamma_1}\frac{\delta p}{p} = \frac{\rho}{\Gamma_1}\frac{\Delta p}{p} = \frac{\rho}{\Gamma_1}\frac{d\ln p}{dz}\Delta z. \qquad (4.3)$$

The change in density of the surroundings is

$$\Delta\rho = \frac{d\rho}{dz}\Delta z. \qquad (4.4)$$

If the element is lighter than its surroundings at height $z + \Delta z$ it will continue to move up. If this is the case, the layer at height $z$ is convectively unstable. The condition for instability is

$$-\delta\rho > -\Delta\rho. \qquad (4.5)$$

Using the above expressions for the changes in density, this becomes

$$-\frac{\rho}{\Gamma_1}\frac{d\ln p}{dz}\Delta z > -\frac{d\rho}{dz}\Delta z. \qquad (4.6)$$

(Note that both sides of the above inequality are positive in radiative layers.) Hence the layer is convectively unstable if

$$\frac{d\ln p}{d\ln\rho} > \Gamma_1, \qquad (4.7)$$

where

$$\frac{d\ln p}{d\ln\rho} = \frac{-d\ln p/dz}{-d\ln\rho/dz}. \qquad (4.8)$$

The inequality (4.7) is the *Schwarzschild criterion* for convection. Because the radiative flux is related to the temperature gradient, it is useful to re-write the Schwarzschild condition in terms of the temperature gradient. We can use the property of pressure balance to transform to the temperature gradients:

$$\delta p = \left.\frac{\partial p}{\partial \rho}\right|_T \delta\rho + \left.\frac{\partial p}{\partial T}\right|_\rho \delta T = \Delta p = \left.\frac{\partial p}{\partial \rho}\right|_T \Delta\rho + \left.\frac{\partial p}{\partial T}\right|_\rho \Delta T. \tag{4.9}$$

Re-arranging, we find

$$\left.\frac{\partial p}{\partial \rho}\right|_T (\delta\rho - \Delta\rho) = -\left.\frac{\partial p}{\partial T}\right|_\rho (\delta T - \Delta T). \tag{4.10}$$

In most circumstances $\left.\frac{\partial p}{\partial \rho}\right|_T$, $\left.\frac{\partial p}{\partial T}\right|_\rho$ are positive. Hence the instability condition

$$\delta\rho - \Delta\rho < 0, \tag{4.11}$$

becomes

$$\delta T - \Delta T > 0, \tag{4.12}$$

i.e. the element must be hotter than its surroundings to continue rising. Since $\delta p = \Delta p < 0$, this gives

$$\frac{\delta T}{\delta p} - \frac{\Delta T}{\Delta p} < 0. \tag{4.13}$$

Since the element is moved adiabatically,

$$\frac{\delta T}{\delta p} = \left.\frac{\partial \ln T}{\partial \ln p}\right|_S \frac{T}{p} = \nabla_{ad}\frac{T}{p}, \tag{4.14}$$

where $\nabla_{ad}$ is the *adiabatic gradient*.

Also since

$$\Delta p = \frac{dp}{dz}\Delta z, \qquad \Delta T = \frac{dT}{dz}\Delta z, \tag{4.15}$$

the Schwarzschild criterion for convective instability can be written as

$$\nabla \equiv \frac{d \ln T}{d \ln p} > \nabla_{ad}, \tag{4.16}$$

where $\nabla$ is the structural gradient.

Note that in deriving the form of the Schwarzschild criterion in terms of temperature gradients, we have assumed that the star is chemically homogeneous. There is some debate as to the correct criterion for convective instability in the presence of composition gradients. However when radiative energy transfer is included in its derivation, the condition (4.16) for convective instability remains correct even in the presence of composition gradients.

## 4.3 Including convective energy transport in stellar models

If a region of a star is convective, then there is an additional 'channel' for transporting energy, which adds to the radiative energy flux. If all the energy were transported radiatively, the temperature gradient would be such that

$$\frac{\mathrm{d}\ln T}{\mathrm{d}\ln p} = \frac{\mathrm{d}\ln T/\mathrm{d}r}{\mathrm{d}\ln p/\mathrm{d}r} = \frac{-\dfrac{3\kappa\rho L}{16\pi a c r^2 T^4}}{-\dfrac{Gm\rho}{pr^2}} = \frac{3\kappa pL}{16\pi a c Gm T^4} \equiv \nabla_{\mathrm{rad}}. \qquad (4.17)$$

This expression is called the *radiative gradient*. Because it carries part of the energy flux, convection acts to reduce the structural gradient below the radiative gradient. Hence in convective regions, the ordering of the three gradients is

$$\nabla_{\mathrm{rad}} \geqslant \nabla \geqslant \nabla_{\mathrm{ad}}. \qquad (4.18)$$

If convection is very efficient, then $\nabla$ becomes very nearly equal to $\nabla_{\mathrm{ad}}$. MS stars more massive than the Sun have convective cores. In these convective cores, the density is high so the thermal content of the convective elements is large and convection is efficient. If convection is inefficient, then very little energy is transported by convection and so $\nabla \approx \nabla_{\mathrm{rad}}$.

To model intermediate conditions, which occur for example in the outer layers of the Sun, a simple phenomenological mixing length theory is used [1, 2]. The mixing length can be thought of as the characteristic size of the convective cells or the average distance a convective element moves vertically before dissolving into the background and depositing its heat.

Let the mixing length be $l$. After moving a mixing length in the vertical direction, the convective element is hotter than its surroundings by

$$\delta T - \Delta T = lT\left(\frac{\mathrm{d}\ln T}{\mathrm{d}z}\bigg|_{\mathrm{ad}} - \frac{\mathrm{d}\ln T}{\mathrm{d}z}\right) = lT\left(\frac{\mathrm{d}\ln T}{\mathrm{d}\ln p}\bigg|_{\mathrm{ad}} - \frac{\mathrm{d}\ln T}{\mathrm{d}\ln p}\right)\frac{\mathrm{d}\ln p}{\mathrm{d}z} = lT(\nabla - \nabla_{\mathrm{ad}})\frac{\rho g}{p}.$$
$$(4.19)$$

The convective energy flux is

$$F_{\mathrm{conv}} \approx v\rho C_p(\delta T - \Delta T), \qquad (4.20)$$

where $v$ is the convective velocity, and $C_p$ is the specific heat at constant pressure.

To estimate $v$, we consider the buoyancy force acting on the convective element. As the convective element rises, it is lighter than its surroundings by

$$-\left(\delta\rho - \Delta\rho\right) = \frac{\partial p/\partial T|_\rho}{\partial p/\partial\rho|_T}(\delta T - \Delta T) = QT(\nabla - \nabla_{\mathrm{ad}})\frac{\rho g}{p}\Delta z, \qquad (4.21)$$

where $\Delta z$ is the distance the convective element has traveled from its starting point, and $Q$ is a (positive) thermal expansion coefficient. Hence the equation of motion for the convective element is

$$\rho\frac{d^2\Delta z}{dt^2} = -\left(\delta\rho - \Delta\rho\right)g = QT(\nabla - \nabla_{ad})\frac{\rho g^2}{p}\Delta z. \tag{4.22}$$

Assuming that the mixing length is sufficiently small that everything except $\Delta z$ in this equation can be treated as constant, we can convert it into a conservation of energy equation by multiplying by $d\Delta z/dt$, and integrating with respect to time to obtain

$$v^2 = QT(\nabla - \nabla_{ad})\frac{g^2}{p}l^2. \tag{4.23}$$

Putting the pieces together, we find that convective flux is given by

$$F_{conv} = Q^{1/2}C_p\left(\frac{T}{p}\right)^{3/2}(\nabla - \nabla_{ad})^{3/2}\rho^2g^2l^2. \tag{4.24}$$

This can be added to the radiative flux to give an equation for the structural gradient. The total luminosity carried by radiation and convection is

$$L = 4\pi r^2(F_{rad} + F_{conv}) = -\frac{16\pi acr^2T^3}{3\kappa\rho}\frac{dT}{dr} + 4\pi r^2Q^{1/2}C_p\left(\frac{T}{p}\right)^{3/2}(\nabla - \nabla_{ad})^{3/2}\rho^2g^2l^2$$

$$= \frac{16\pi acGmT^4}{3\kappa p}\nabla + 4\pi r^2Q^{1/2}C_p\left(\frac{T}{p}\right)^{3/2}(\nabla - \nabla_{ad})^{3/2}\rho^2g^2l^2. \tag{4.25}$$

The luminosity can be written in terms of the radiative gradient,

$$L = \frac{16\pi acGmT^4}{3\kappa p}\nabla_{rad}, \tag{4.26}$$

to obtain

$$\nabla_{rad} = \nabla + A(\nabla - \nabla_{ad})^{3/2}, \tag{4.27}$$

where

$$A = \frac{3\kappa p}{4acT^4}\rho^2gl^2Q^{1/2}C_p\left(\frac{T}{p}\right)^{3/2}. \tag{4.28}$$

It is convenient to introduce the quantity

$$\Gamma = \frac{\nabla_{rad} - \nabla}{\nabla - \nabla_{ad}}, \tag{4.29}$$

which ranges from 0 to infinity depending on the efficiency of convection. In terms of this quantity

$$\nabla = \frac{\nabla_{\text{rad}} + \Gamma\nabla_{\text{ad}}}{1 + \Gamma}, \tag{4.30}$$

and

$$\Gamma^3 + \Gamma^2 = A^2(\nabla_{\text{rad}} - \nabla_{\text{ad}}), \tag{4.31}$$

which is easily solved for $\Gamma$.

This is the simplest form of mixing length theory. More sophisticated forms exist in which, for example, the convective element loses its excess energy as it moves rather than just at the end of its life. All that remains is to specify the mixing length, $l$. This is usually taken to be proportional to the pressure scale height

$$l = \alpha H_p = \alpha \left| \frac{\text{d}r}{\text{d}\ln p} \right| = \alpha\frac{p}{\rho g}. \tag{4.32}$$

The mixing length ratio, $\alpha$, is fixed by calibrating with the properties of the Sun. It is often found to be about 1.5–2.0 (there is not a unique value for $\alpha$ because differences in treatment of the 'physics' of the Sun can be compensated to some extent by changing $\alpha$). Note that 1.5–2.0 is not small compared to unity and hence the assumption that the properties of the surroundings are approximately constant over the mixing length is not strictly valid. This is one of the many deficiencies of mixing length theory.

## Bibliography

[1] Prandtl L 1925 Z. Angew. Math. Mech. **5** 136
[2] Böhm-Vitense E 1958 Z. Astrophys. **46** 108

Structure and Evolution of Single Stars
An introduction
**James MacDonald**

# Chapter 5

## The equations of stellar evolution
## and how to solve them

### 5.1 Introduction

Here we review the equations of stellar structure and consider appropriate boundary conditions. Equations that describe the composition changes are introduced. Finally the most common method for solving the complete set of equations is outlined.

### 5.2 The equations of stellar structure

These are the continuity equation

$$\frac{dm}{dr} = 4\pi r^2 \rho, \tag{5.1}$$

the hydrostatic balance equation

$$\frac{dp}{dr} = -\rho \frac{Gm}{r^2}, \tag{5.2}$$

the conservation of energy equation

$$\frac{dL}{dm} = \varepsilon - \frac{du}{dt} + \frac{p}{\rho^2} \frac{d\rho}{dt}, \tag{5.3}$$

and the energy transport equation

$$\frac{d \ln T}{d \ln p} = \frac{\nabla_{rad} + \Gamma \nabla_{ad}}{1 + \Gamma}, \tag{5.4}$$

where

$$\Gamma = 0, \quad \text{if } \nabla_{rad} \leqslant \nabla_{ad},$$
$$\Gamma^3 + \Gamma^2 = A^2 (\nabla_{rad} - \nabla_{ad}), \quad \text{if } \nabla_{rad} > \nabla_{ad}. \tag{5.5}$$

doi:10.1088/978-1-6817-4105-5ch5

In the energy transport equation

$$\nabla_{\text{rad}} = \frac{3\kappa p L}{16\pi ac GmT^4},$$ (5.6)

and

$$A = \frac{3\kappa p}{4ac T^4}\rho^2 gl^2 Q^{1/2}C_p\left(\frac{T}{p}\right)^{3/2}.$$ (5.7)

Once we find expressions for $p$, $u$, $\varepsilon$, $\nabla_{\text{ad}}$, $\kappa$, $Q$ and $C_p$ in terms of $\rho$, $T$ and the composition variables, we have a complete set of equations. Since there are four first order differential equations, we need to specify four boundary conditions to obtain a complete solution. Two of the boundary conditions are central boundary conditions and the other two are surface boundary conditions. The central boundary conditions are simply

$$\left.\begin{array}{l} r = 0 \\ L = 0 \end{array}\right\} \text{at } m = 0.$$ (5.8)

The surface boundary conditions are not so straightforward. At first sight it would appear that they should be $p = 0$, $T = 0$ at the surface (where $m = M$). However these conditions are not used for a number of reasons, including (i) the radiative energy transport equation is not valid at low optical depth, (ii) there is a non-zero pressure associated with the radiation that escapes from the star, and (iii) the useful variables $\ln T$, $\ln p$ would be singular at the surface. To avoid these problems, the boundary conditions are applied a little way in from the surface. A particularly useful location is the photosphere at optical depth, $\tau = 2/3$. There the temperature is related to the radius and luminosity by

$$L = \pi acr^2 T^4.$$ (5.9)

The second boundary condition is obtained from the hydrostatic balance equation

$$\frac{\mathrm{d}p}{\mathrm{d}r} = -g\rho,$$ (5.10)

and the equation for the optical depth

$$\frac{\mathrm{d}\tau}{\mathrm{d}r} = -\kappa\rho.$$ (5.11)

These give that

$$\frac{\mathrm{d}p}{\mathrm{d}\tau} = \frac{g}{\kappa}.$$ (5.12)

Because the radiation pressure does not go to zero at $\tau = 0$, we separate the pressure into its contributions from gas and radiation,

$$\frac{dp_{gas}}{d\tau} = \frac{g}{\kappa} - \frac{dp_{rad}}{d\tau} = \frac{g}{\kappa} - \frac{4aT^3}{3}\frac{dT}{d\tau}. \tag{5.13}$$

Assuming that the photosphere is in radiative equilibrium, using the diffusion equation we find

$$\frac{dp_{gas}}{d\tau} = \frac{g}{\kappa}\left(1 - \frac{L}{L_{Ed}}\right), \tag{5.14}$$

where $L_{Ed}$ is called the *Eddington luminosity*,

$$L_{Ed} = \frac{4\pi cGM}{\kappa}. \tag{5.15}$$

We now assume that everything on the right-hand side of equation (5.14) is constant so that

$$p_{gas} = \frac{g}{\kappa}\left(1 - \frac{L}{L_{Ed}}\right)\tau. \tag{5.16}$$

We have further assumed that density is zero at optical depth zero. At the photosphere

$$p_{gas}\left(\tau = \frac{2}{3}\right) = \frac{2}{3}\frac{g}{\kappa}\left(1 - \frac{L}{L_{Ed}}\right). \tag{5.17}$$

This is the second surface boundary condition.

The boundary conditions given in equations (5.9) and (5.17) are used mainly because of their simplicity and because the approximations used in deriving equation (5.17) cannot be expected to be completely accurate. An alternative approach is to use surface boundary conditions obtained from detailed stellar atmosphere models (see e.g. [1]).

## 5.3 The physical significance of the Eddington luminosity

Since pressure must be positive, the surface boundary conditions can be satisfied only if $L < L_{Ed}$. The surface Eddington luminosity is an upper limit on the luminosity of a (spherical) star for it to be in hydrostatic equilibrium. Inside the star, if the luminosity exceeds the Eddington limit in some region, then (i) that region either is or becomes convective, (ii) there is a density inversion so that the gradient in gas pressure opposes the radiative force, or (iii) there is a loss of hydrostatic equilibrium.

## 5.4 Equations for composition changes

The composition of stellar material changes due to a number of physical processes including (i) nuclear transformations and (ii) turbulent convective mixing.

As we shall see later when we consider the nuclear reactions in detail, there are a variety of nuclear processes that can transmute one nuclear species to another. In general the change in mass fraction of a nuclear species $k$ is described by an equation of form

$$\frac{\mathrm{d}X_k}{\mathrm{d}t} = C_k(\mathbf{X}, \rho, T) - D_k(\mathbf{X}, \rho, T)X_k, \tag{5.18}$$

where $C_k$ is the rate at which species $k$ is created and $D_kX_k$ is the rate at which it is destroyed.

Since turbulent convective mixing acts to smooth out composition gradients, it is often modeled as a diffusion process. In this case, a diffusion term that describes the convective mixing is added to the equation above

$$\frac{\mathrm{d}X_k}{\mathrm{d}t} = C_k(\mathbf{X}, \rho, T) - D_k(\mathbf{X}, \rho, T)X_k + \nabla \cdot \left(\sigma_c \nabla X_k\right), \tag{5.19}$$

where, in the mixing length model, the turbulent diffusivity is

$$\sigma_c = \beta vl. \tag{5.20}$$

Here $v$ and $l$ are the convective velocity and mixing length, and $\beta$ is a constant of order unity. In convectively stable regions, $\sigma_c = 0$.

## 5.5 Solving the equations of stellar evolution

Except for a few simple models, the equations of stellar evolution cannot be solved analytically. A numerical approach is needed. Furthermore, because the equations are time-dependent, initial conditions must be specified. These include specifying the model star's mass $M$ and initial composition (usually assumed to be uniform).

The equations of stellar evolution are solved by first approximating them by *finite difference equations*. The star is divided up into a set of nested concentric spherical shells. Suppose the density in the $k$th such shell is $\rho_k$. Then, for example, the continuity equation can be finite differenced as

$$m_{k+1} - m_k = \frac{4\pi}{3}\rho_k\left(r_{k+1}^3 - r_k^3\right). \tag{5.21}$$

Here $m_k, r_k, m_{k+1}, r_{k+1}$ are the values of $m$ and $r$ at the boundaries of the shell (see figure 5.1). A similar finite difference procedure is used for the time derivatives, e.g.

$$\frac{\mathrm{d}u_k}{\mathrm{d}t} = \frac{u_k^{n+1} - u_k^n}{\Delta t}, \tag{5.22}$$

where $\Delta t$ is the time step, and $u_k^n$ is the value of $u$ in the $k$th shell at the $n$th time step. Assuming the mass coordinates are fixed, the continuity equation advanced over a time step is

$$m_{k+1} - m_k = \frac{4\pi}{3}(\rho_k + \Delta\rho_k)\left[\left(r_{k+1} + \Delta r_{k+1}\right)^3 - \left(r_k + \Delta r_k\right)^3\right]. \tag{5.23}$$

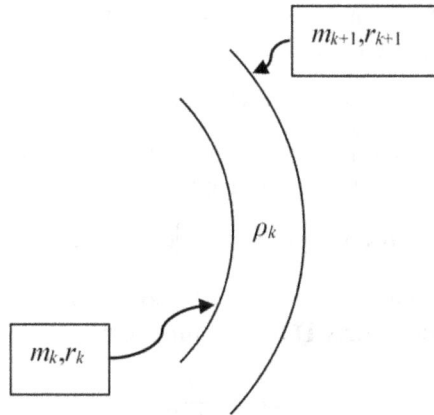

**Figure 5.1.** Schematic for finite differencing of the mass conservation equation.

When combined with the finite difference forms of all the other stellar evolution equations, we have a set of coupled non-linear algebraic equations for the increments $\Delta \rho_k$, $\Delta r_k$, $\Delta r_{k+1}$, etc. These are solved iteratively using the Newton–Raphson method.

## 5.6 The Newton–Raphson method

Suppose we have a guess, $x_n$, for a solution of the equation

$$f(x) = 0. \tag{5.24}$$

Denote the difference between the exact solution and the guess by $\Delta x$, so that

$$f(x_n + \Delta x) = 0. \tag{5.25}$$

Expanding the left-hand side in a Taylor series gives

$$f(x_n) + f'(x_n)\Delta x + \cdots = 0. \tag{5.26}$$

By truncating the left-hand side after the second term, we have

$$\Delta x \simeq -\frac{f(x_n)}{f'(x_n)}. \tag{5.27}$$

A new estimate for the solution is obtained by adding the right-hand side to $x_n$,

$$x_{n+1} = x_n - \frac{f(x_n)}{f'(x_n)}. \tag{5.28}$$

This is a recurrence relation that can be used repeatedly until the desired accuracy is obtained. If the initial guess is close enough to the solution, convergence is usually very rapid.

## 5.7 Sets of non-linear equations

Because the concept of bracketing cannot be generalized to higher dimensions, the only general method to solve a set of non-linear equations, such as the equations of

5-5

stellar evolution, is the Newton–Raphson method. We can write the set of equations in a compact form using vector notation

$$\mathbf{f}(\mathbf{x}) = 0, \tag{5.29}$$

where $\mathbf{x}$ is a vector of the quantities to be solved for and $\mathbf{f}$ is a vector of the equations relating these quantities. The truncated Taylor series expansion is

$$\mathbf{f}(\mathbf{x}_n) + \left[ (\Delta\mathbf{x} \cdot \nabla)\mathbf{f} \right](\mathbf{x}_n) = 0. \tag{5.30}$$

This is a set of linear equations that can be solved by matrix methods. To make this clearer, introduce the matrix $\mathbf{D}$ with components

$$D_{ij} = \frac{\partial f_i}{\partial x_j}. \tag{5.31}$$

The components of $\Delta\mathbf{x}$ satisfy

$$\sum_j D_{ij}\Delta x_j = b_i, \tag{5.32}$$

where

$$b_i = -f_i(\mathbf{x}). \tag{5.33}$$

More detailed descriptions of how the stellar evolution equations are solved numerically can be found in many books, including [2] and [3].

## Bibliography

[1] Baraffe I, Chabrier G, Allard F and Hauschildt P H 1997 *Astron. Astrophys.* **327** 1054
[2] Kippenhahn R, Weigert A and Weiss A 2012 *Stellar Structure and Evolution* 2nd edn (Berlin: Springer) doi:10.1007/978-3-642-30304-3
[3] Bodenheimer P, Laughlin G P, Rozyczka M and Yorke H W 2006 *Numerical Methods in Astrophysics: An Introduction* (Boca Raton, FL: CRC)

# Chapter 6

## Physics of gas and radiation

## 6.1 Introduction

We begin consideration of the equation of state of stellar material. By equation of
state is meant the relationships between the various thermodynamic variables. These
include the extensive variables, internal energy $U$, entropy $S$, and volume $V$, and the
intensive variables, temperature $T$ and pressure $p$. A methodical approach to
deriving the equation of state is by modeling the Helmholtz free energy, $F$, of a
system of interacting particles in a fixed volume at fixed temperature [1, 2].
Thermodynamic equilibrium of the system requires $F$ to take a minimum value
under stoichiometric constraints, which are determined from possible dissociation
and ionization processes, and constraints from overall charge neutrality and particle
number conservation. The Helmholtz free energy is related to the internal energy of
the system, its entropy, and temperature by

$$F = U - TS. \tag{6.1}$$

From the laws of thermodynamics, for any small change in the system (in
equilibrium)

$$dU = T dS - p dV. \tag{6.2}$$

Hence

$$dF = dU - T dS - S dT = -p dV - S dT. \tag{6.3}$$

Since $dF$ is an exact differential, we must have

$$p = -\left.\frac{\partial F}{\partial V}\right|_T, \quad S = -\left.\frac{\partial F}{\partial T}\right|_V, \quad U = F - T\left.\frac{\partial F}{\partial T}\right|_V = -T^2 \frac{\partial}{\partial T}\left(\frac{F}{T}\right)\bigg|_V. \tag{6.4}$$

Hence $p$, $S$, and $U$, are readily obtained if $F$ for the system in thermodynamic
equilibrium is known in terms of $V$ and $T$.

## 6.2 The ideal gas equation of state

To find the entropy of an ideal gas, we do not need to find $F$ first. We can make use of the expressions for the pressure and internal energy for $N$ particles in volume $V$. The pressure is

$$p = \frac{N}{V}kT, \tag{6.5}$$

and, for point particles with no internal degrees of freedom, the total internal energy is

$$U = \frac{3}{2}NkT. \tag{6.6}$$

Since

$$dU = T\,dS - p\,dV, \tag{6.7}$$

we have

$$\frac{3}{2}Nk\,dT = T\,dS - \frac{N}{V}kT\,dV. \tag{6.8}$$

After some re-arrangement, we find

$$dS = \frac{3}{2}Nk\,d\ln T + Nk\,d\ln V = Nk\,d\ln(T^{3/2}V), \tag{6.9}$$

so that

$$S = Nk\,\ln(T^{3/2}V) + C(N), \tag{6.10}$$

where $C(N)$ is independent of $V$ and $T$. Since $S$ is an extensive quantity, we must have

$$S = Nk\,\ln\left(c\frac{V}{N}T^{3/2}\right), \tag{6.11}$$

where $c$ is a constant that cannot be determined from thermodynamics alone.

We see immediately that for an adiabatic change (i.e. one in which $S$ is constant),

$$\frac{\delta V}{V} + \frac{3}{2}\frac{\delta T}{T} = 0. \tag{6.12}$$

Hence

$$\frac{\delta p}{p} = -\frac{\delta V}{V} + \frac{\delta T}{T} = \frac{5}{2}\frac{\delta T}{T} = -\frac{5}{3}\frac{\delta V}{V}. \tag{6.13}$$

Since

$$\frac{\delta \rho}{\rho} = -\frac{\delta V}{V}, \tag{6.14}$$

it follows that for an ideal gas

$$\Gamma_1 = \frac{5}{3}, \quad \nabla_{\text{ad}} = \frac{2}{5}. \tag{6.15}$$

For mixing length theory, we also need an expression for the specific heat. The specific heat at constant pressure is defined by

$$C_p = T \left. \frac{\partial s}{\partial T} \right|_p, \tag{6.16}$$

where $s$ is the entropy per unit mass. Since the mean mass per particle is $\mu m_{\text{u}}$,

$$s = \frac{S}{N \mu m_{\text{u}}} = \frac{k}{\mu m_{\text{u}}} \ln \left( c \frac{V}{N} T^{3/2} \right) = \frac{\mathfrak{R}}{\mu} \ln \left( ck \frac{T^{5/2}}{p} \right). \tag{6.17}$$

The specific heat at constant pressure is then

$$C_p = \frac{5}{2} \frac{\mathfrak{R}}{\mu}. \tag{6.18}$$

Similarly, the specific heat at constant volume is

$$C_V = T \left. \frac{\partial s}{\partial T} \right|_V = \frac{3}{2} \frac{\mathfrak{R}}{\mu}. \tag{6.19}$$

Note that for an ideal gas the ratio of the specific heats is equal to the first adiabatic exponent. However this is not a general result.

## 6.3 The radiation equation of state

We begin by deriving a relationship between the pressure and energy density for fully relativistic particles.

For a particle of rest mass $m$ moving with speed $v$, its energy and momentum are

$$\begin{aligned} E &= \gamma m c^2, \\ q &= \gamma m v, \end{aligned} \tag{6.20}$$

where the Lorentz factor

$$\gamma = \left( 1 - \frac{v^2}{c^2} \right)^{-1/2}. \tag{6.21}$$

(We use $q$ for momentum rather than the traditional $p$, because we are using $p$ as the symbol for pressure.)

On eliminating $v$ and $\gamma$ from these expressions, we obtain

$$E^2 = m^2 c^4 + c^2 q^2. \tag{6.22}$$

We will use this expression later when we consider the equation of state for degenerate electrons. Here we are interested in massless photons, for which

$$E = qc. \tag{6.23}$$

The energy per unit volume is

$$\varepsilon = \int_0^\infty E(q)n(q)4\pi q^2 dq, \tag{6.24}$$

where $n(q)$ is the momentum distribution function, i.e. the number of photons per unit volume with momentum between $q$ and $q + dq$ is $n(q)dq$.

To evaluate the pressure consider a planar surface with the normal in the x-direction. The pressure is related to the momentum flux through this surface, so that

$$p = \int_0^\infty q_x v_x n(q)4\pi q^2 dq. \tag{6.25}$$

Since $q_x$ and $v_x$ are parallel, we have

$$v_x = \frac{q_x}{q}c, \tag{6.26}$$

which gives

$$p = \int_0^\infty q_x^2 cn(q)4\pi q dq. \tag{6.27}$$

Assuming that there are no preferred directions, the choice of the x-direction is arbitrary. Hence

$$3p = \int_0^\infty \left(q_x^2 + q_y^2 + q_y^2\right)cn(q)4\pi q dq = \int_0^\infty q^2 cn(q)4\pi q dq$$
$$= \int_0^\infty E(q)n(q)4\pi q^2 dq = \varepsilon. \tag{6.28}$$

For isotropic radiation, we find

$$p = \frac{1}{3}\varepsilon. \tag{6.29}$$

We can now find $p$ in terms of temperature by making some manipulations of thermodynamic relations. The exact differential of the entropy is

$$dS = \frac{dU}{T} + \frac{pdV}{T}. \tag{6.30}$$

Consider $S$ and $U$ as functions of $V$ and $T$, so that

$$dS = \frac{\partial S}{\partial T}dT + \frac{\partial S}{\partial V}dV = \frac{1}{T}\frac{\partial U}{\partial T}dT + \frac{1}{T}\frac{\partial U}{\partial V}dV + \frac{pdV}{T}. \tag{6.31}$$

since $T$ and $V$ can be varied independently, we must have

$$\frac{\partial S}{\partial T} = \frac{1}{T}\frac{\partial U}{\partial T}, \tag{6.32}$$

and

$$\frac{\partial S}{\partial V} = \frac{1}{T}\frac{\partial U}{\partial V} + \frac{p}{T}. \tag{6.33}$$

The integrability relation for $dS$ to be an exact differential is

$$\frac{\partial}{\partial V}\left(\frac{\partial S}{\partial T}\right) = \frac{\partial}{\partial T}\left(\frac{\partial S}{\partial V}\right), \tag{6.34}$$

which gives

$$\frac{1}{T}\frac{\partial^2 U}{\partial V \partial T} = \frac{\partial}{\partial T}\left(\frac{1}{T}\frac{\partial U}{\partial V} + \frac{p}{T}\right) = \frac{1}{T}\frac{\partial^2 U}{\partial T \partial V} - \frac{1}{T^2}\frac{\partial U}{\partial V} + \frac{\partial}{\partial T}\left(\frac{p}{T}\right). \tag{6.35}$$

Since $dU$ is also an exact differential, this gives

$$\frac{1}{T^2}\frac{\partial U}{\partial V} = \frac{\partial}{\partial T}\left(\frac{p}{T}\right). \tag{6.36}$$

For a volume $V$ of radiation the total energy is $U = V\varepsilon(T)$ and pressure is a function of $T$ alone. Hence

$$\frac{d}{dT}\left(\frac{p}{T}\right) = \frac{1}{T^2}\varepsilon = 3\frac{p}{T^2}. \tag{6.37}$$

Integrating with respect to $T$ gives

$$p = \frac{1}{3}aT^4, \tag{6.38}$$

where $a$ is a constant of integration (the radiation constant). The internal energy per unit volume is

$$\varepsilon = aT^4, \tag{6.39}$$

and the entropy per unit volume is

$$\frac{S}{V} = \frac{4}{3}aT^3. \tag{6.40}$$

## 6.4 The equation of state for a mixture of ideal gas and radiation

Combining the results above for ideal gases and radiation, we have pressure

$$p = \frac{\Re}{\mu}\rho T + \frac{1}{3}aT^4, \tag{6.41}$$

and internal energy per unit mass

$$u = \frac{3}{2}\frac{\Re}{\mu}T + \frac{aT^4}{\rho}. \tag{6.42}$$

Since for unit mass $V = 1/\rho$, we have for an adiabatic change,

$$\mathrm{d}u = \frac{p}{\rho^2}\mathrm{d}\rho. \tag{6.43}$$

Using the above expressions for $u$ and $p$, we obtain

$$\frac{3}{2}\frac{\Re}{\mu}\mathrm{d}T + 4\frac{aT^3}{\rho}\mathrm{d}T - \frac{aT^4}{\rho^2}\mathrm{d}\rho = \frac{T}{\rho}\frac{\Re}{\mu}\mathrm{d}\rho + \frac{1}{3}\frac{aT^4}{\rho^2}\mathrm{d}\rho. \tag{6.44}$$

Hence

$$\left(\frac{3}{2}\frac{\Re}{\mu}\rho T + 4aT^4\right)\frac{\mathrm{d}T}{T} = \left(\frac{\Re}{\mu}\rho T + \frac{4}{3}aT^4\right)\frac{\mathrm{d}\rho}{\rho}. \tag{6.45}$$

It is convenient to introduce the ratio of the ideal gas pressure to the total pressure

$$\beta = \frac{p_{\text{gas}}}{p}, \tag{6.46}$$

so that

$$1 - \beta = \frac{p_{\text{rad}}}{p} = \frac{1}{3}\frac{aT^4}{p}. \tag{6.47}$$

In terms of $\beta$,

$$(24 - 21\beta)\frac{\mathrm{d}T}{T} = (8 - 6\beta)\frac{\mathrm{d}\rho}{\rho}. \tag{6.48}$$

To determine whether convection occurs, we need

$$\Gamma_1 = \frac{\partial \ln p}{\partial \ln \rho}\bigg|_s,$$

and

$$\nabla_{\text{ad}} = \frac{\partial \ln T}{\partial \ln p}\bigg|_s.$$

Since

$$\frac{\mathrm{d}p}{p} = \frac{\Re}{\mu}\frac{\rho T}{p}\frac{\mathrm{d}\rho}{\rho} + \frac{\Re}{\mu}\frac{\rho T}{p}\frac{\mathrm{d}T}{T} + \frac{4}{3}\frac{aT^4}{p}\frac{\mathrm{d}T}{T}$$

$$= \frac{p_{\text{gas}}}{p}\frac{\mathrm{d}\rho}{\rho} + \frac{p_{\text{gas}}}{p}\frac{\mathrm{d}T}{T} + 4\frac{p_{\text{rad}}}{p}\frac{\mathrm{d}T}{T}$$

$$= \beta\frac{\mathrm{d}\rho}{\rho} + (4 - 3\beta)\frac{\mathrm{d}T}{T}, \tag{6.49}$$

we obtain

$$\Gamma_1 = \left.\frac{\partial \ln p}{\partial \ln \rho}\right|_s = \frac{\beta\dfrac{\mathrm{d}\rho}{\rho} + (4 - 3\beta)\dfrac{\mathrm{d}T}{T}}{\dfrac{\mathrm{d}\rho}{\rho}} = \beta + \frac{(4 - 3\beta)(8 - 6\beta)}{(24 - 21\beta)}$$

$$= \frac{32 - 24\beta - 3\beta^2}{24 - 21\beta}. \tag{6.50}$$

Similarly, we find

$$\nabla_{\text{ad}} = \left.\frac{\partial \ln T}{\partial \ln p}\right|_s = \frac{8 - 6\beta}{32 - 24\beta - 3\beta^2}. \tag{6.51}$$

Hence $\Gamma_1$ ranges from 5/3 when radiation pressure is negligible to 4/3 when radiation pressure dominates. Also $\nabla_{\text{ad}}$ ranges from 0.4 to 0.25. We see that the presence of radiation decreases both $\Gamma_1$ and $\nabla_{\text{ad}}$.

This approach can be extended to find

$$C_p = T\left.\frac{\partial s}{\partial T}\right|_p = \frac{\Re}{\mu}\frac{1}{2\beta^2}\left(32 - 24\beta - 3\beta^2\right), \tag{6.52}$$

and

$$Q = \frac{\partial p/\partial T|_\rho}{\partial p/\partial \rho|_T} = \frac{\rho}{T}\frac{\partial \ln p/\partial \ln T|_\rho}{\partial \ln p/\partial \ln \rho|_T} = \frac{\rho}{T}\frac{4 - 3\beta}{\beta}. \tag{6.53}$$

## 6.5 The Eddington standard model of stellar structure

Suppose that $\beta$ is uniform throughout the star, and that the star is radiative everywhere. Since $p_{\text{rad}} = (1 - \beta)p$,

$$\frac{\mathrm{d}\ln p_{\text{rad}}}{\mathrm{d}\ln p} = 4\frac{\mathrm{d}\ln T}{\mathrm{d}\ln p} = 1. \tag{6.54}$$

Hence $\nabla = 1/4$ everywhere in the star. Since the star is radiative,

$$\nabla_{\text{rad}} = \frac{3\kappa p L}{16\pi ac GmT^4} = \frac{\kappa p L}{16\pi c Gmp_{\text{rad}}} = \frac{\kappa L}{16\pi c Gm(1-\beta)} = \frac{1}{4}. \qquad (6.55)$$

We conclude that $\beta$ can be uniform throughout the star only if $\kappa L/m$ is uniform throughout the star. This is the assumption behind Eddington's standard model [3].

Furthermore, since

$$P_{\text{gas}} = \frac{\Re}{\mu}\rho T = \beta p, \qquad P_{\text{rad}} = \frac{1}{3}aT^4 = (1-\beta)p, \qquad (6.56)$$

we have on eliminating $T$,

$$p = \left[\frac{(1-\beta)}{\beta^4}\frac{3}{a}\left(\frac{\Re}{\mu}\right)^4\right]^{1/3}\rho^{4/3}. \qquad (6.57)$$

This is an example of what is called a *polytropic* relation between pressure and density. We shall return to polytropic models when we consider the structure of WD stars. Here we note that simple scaling arguments applied to the continuity and hydrostatic balance equations give

$$p_c \sim \frac{GM^2}{R^4}, \qquad \rho_c \sim \frac{M}{R^3}, \qquad (6.58)$$

so that

$$\frac{GM^2}{R^4} \sim \left[\frac{(1-\beta)}{\beta^4}\frac{3}{a}\left(\frac{\Re}{\mu}\right)^4\right]^{1/3}\left(\frac{M}{R^3}\right)^{4/3}. \qquad (6.59)$$

Note that the radius drops out of this relation, and we find[1]

$$M^2 \sim \frac{(1-\beta)}{\beta^4}\frac{3}{aG^3}\left(\frac{\Re}{\mu}\right)^4. \qquad (6.60)$$

Hence in Eddington's standard model of stellar structure $\beta$ depends only on the star's mass and its composition through the mean molecular weight. For stars of the same composition, $\beta$ is lower in stars of higher mass, which is consistent with our earlier conclusion that radiation pressure is more important in massive MS stars.

## Bibliography

[1] Harris G M 1959 *J. Chem. Phys.* **31** 1211
[2] Graboske H C *et al* 1969 *Phys. Rev.* **186** 210
[3] Eddington A S 1959 *The Internal Constitution of the Stars* (New York: Dover) p 117

---

[1] The complete relation is $(M/M_\odot)^2 = 331.1(1-\beta)/(\mu\beta)^4$.

# Chapter 7

## Ionization and recombination

### 7.1 Introduction

Photons of sufficient energy can, on absorption, kick an electron off an atom or ion, in a process called *ionization*. The inverse process in which an ion and an electron combine with emission of a photon is called *recombination*. If the atom or ion is in its ground state, the minimum photon energy that can remove an electron is called the *ionization potential*, usually denoted by $I$ or $\chi$. For hydrogen, $\chi_H = 13.6$ eV. Since 1 eV corresponds to a temperature of 11 600 K, $\chi_H$ corresponds to a temperature of $1.5 \times 10^5$ K. Hence it might be expected that temperatures of order $10^5$ K are needed for ionization to occur. However there is an additional factor involved in determining the balance between ionization and recombination. The recombination rate is proportional to the product of the electron number density and the number density of the ions in the higher ionization state, whereas the ionization rate is proportional to the number density of the ions in the lower ionization state. Hence the recombination rate has a quadratic dependence on the density whereas the ionization rate is linearly dependent on the density. A higher temperature is needed for ionization at high density than at low density. At the densities typical of stellar envelopes, there are sufficient numbers of photons in the high energy tail of the Planck distribution for ionization of hydrogen to begin at $T \sim 10\,000$ K, i.e. at a temperature corresponding to about 10% of the ionization potential. Since the ionization potential of neutral helium atoms is $\chi_{He} = 24.6$ eV, ionization of helium atoms begins at $T \sim 20\,000$ K. Similarly, since $\chi_{He^+} = 54.4$ eV, the transition from $He^+$ to $He^{++}$ begins at $T \sim 50\,000$ K.

### 7.2 The Boltzmann excitation equation

Consider a single species of atom which has a number of bound states. Consider a large number of such atoms in contact with a thermal bath at temperature $T$. Let $n_i$ be the number of atoms in state $i$ and let the excitation energy of state $i$ be $\chi_i$. The excitation energy is the energy required to the lift an atom from its ground state to the excited state. Let $g_i$ be the statistical weight of energy level $i$ that accounts for degenerate sublevels. The Boltzmann excitation equation[1] gives that the population of state $i$ relative to the ground state is

$$\frac{n_i}{n_0} = \frac{g_i}{g_0} \exp\left(-\frac{\chi_i}{kT}\right).$$
(7.1)

## 7.3 The Saha ionization equation

Above the discrete bound states, there is a continuum of levels in which an electron is unbound and has non-zero kinetic energy. The energy (measured relative to the ground state) at which this continuum begins is the ionization potential, $\chi$. We can find the relative numbers of atoms and ions in successive stages of ionization from the Saha equation, which we will derive by extending the Boltzmann equation to continuum states.

To start, consider a situation in which an atom in its ground state is ionized, resulting in an ion in its ground state plus a free electron. The energy required to do this is

$$E = \chi + \frac{1}{2}\frac{q^2}{m_e},$$
(7.2)

where $q$ is the electron momentum. From the Boltzmann formula

$$\frac{n_1(q)}{n_0} = \frac{g(q)}{g_0} \exp\left(-\frac{\chi + q^2/(2m_e)}{kT}\right),$$
(7.3)

where $n_1(q)\mathrm{d}q$ is the number density of ions with an accompanying electron that has momentum between $q$ and $q + \mathrm{d}q$, $g(q)\mathrm{d}q$ is the statistical weight of the ion plus electron, $n_0$ is the number density of the atoms, and $g_0$ is the statistical weight of the atom.

The statistical weight of the ion plus electron is the product of the statistical weight of the ion and the statistical weight of the electron

$$g(q) = g_1 g_e(q).$$
(7.4)

To obtain $g_e(q)$, we use Pauli's exclusion principle which states that for fermions, no more than one particle can occupy a quantum state. Here a quantum state is a bin in phase space of volume $h^3$. Since the electron can be in one of two spin states, we obtain

$$g_e(q)\mathrm{d}q = \frac{2}{n_e}\frac{4\pi q^2 \mathrm{d}q}{h^3}.$$
(7.5)

(The $1/n_e$ factor comes from the space volume element. It is the volume per electron.)

---

[1] An interesting geometrical derivation of the Boltzmann factor that does not involve contact with a heat bath is given by [1].

7-2

From equation (7.3) we obtain

$$\frac{n_1(q)}{n_0} = \frac{2g_1}{n_e g_0} \frac{4\pi q^2}{h^3} \exp\left(-\frac{\chi + q^2/(2m_e)}{kT}\right). \tag{7.6}$$

Integrating over $q$ we obtain

$$n_1 = \int_0^\infty n_1(q)\mathrm{d}q = \frac{2n_0 g_1}{n_e g_0} \int_0^\infty 4\pi q^2 \exp\left(-\frac{\chi + q^2/(2m_e)}{kT}\right)\mathrm{d}q$$

$$= \frac{2n_0 g_1}{n_e g_0} \exp\left(-\frac{\chi}{kT}\right)\int_0^\infty 4\pi q^2 \exp\left(-\frac{q^2}{2m_e kT}\right)\mathrm{d}q. \tag{7.7}$$

Evaluating the integral gives

$$\frac{n_1 n_e}{n_0} = 2\frac{g_1}{g_0}\left(\frac{2\pi m_e kT}{h^2}\right)^{3/2} \exp\left(-\frac{\chi}{kT}\right). \tag{7.8}$$

Of course, atoms and ions are not confined to their ground states. To take into account all of the bound states, the statistical weights must be replaced by *partition functions*, which show how the internal states of an ion are populated.

The Saha equation is then

$$\frac{n_1 n_e}{n_0} = 2\frac{G_1}{G_0}\left(\frac{2\pi m_e kT}{h^2}\right)^{3/2} \exp\left(-\frac{\chi}{kT}\right), \tag{7.9}$$

where subscripts 0 and 1 refer to two successive ionization states and the partition functions are given by

$$G = \sum_{\text{bound states}} g_i \exp\left(-\frac{\chi_i}{kT}\right), \tag{7.10}$$

where $\chi_i$ is the energy of an excited state relative to the ground state.

## 7.4 A difficulty and its resolution

Let us apply the Saha equation to the ionization of atomic hydrogen. The energy levels of the hydrogen atom relative to its ground state are

$$\chi_k = \chi_H\left[1 - \frac{1}{(k+1)^2}\right], \tag{7.11}$$

where $n = k + 1$ is the principal quantum number of the state. If we fix the spin of the nuclear proton, the statistical weight of state $n$ is $2n^2$. Hence

$$G = \sum_{n=1}^\infty 2n^2 \exp\left[-\frac{\chi_H}{kT}\left(1 - \frac{1}{n^2}\right)\right] = \exp\left(-\frac{\chi_H}{kT}\right)\sum_{n=1}^\infty 2n^2 \exp\left(\frac{\chi_H}{kT}\frac{1}{n^2}\right). \tag{7.12}$$

Since $\exp(\frac{\chi_H}{kT}\frac{1}{n^2}) > 1$, we see that the partition function diverges!

This is correct for a single isolated hydrogen atom. However hydrogen atoms in a star are not isolated. They interact with other particles. The inter-particle interactions modify the statistical weights of the internal states (and to a lesser extent the energy levels). This can be modeled by introducing *occupation probabilities* [2], $w_i$, such that

$$G = \sum_{\text{bound states}} w_i g_i \exp\left(-\frac{\chi_i}{kT}\right), \qquad (7.13)$$

where $w_i \to 0$, as $i \to \infty$. The $w_i$ depend on the nature of the interactions.

## 7.5 Ionization of hydrogen

Hydrogen can exist in many forms including $H_2$, $H_2^+$, $H^-$, H, and $H^+$. Dissociation of $H_2$ is important in lower MS stars and $H^-$ is an important opacity source in the Sun and solar-like stars. Here we will consider the ionization of H to $H^+$.

To obtain some insight into the conditions at which ionization occurs in a pure hydrogen gas, we will make the approximation $G = g_0$ for atomic hydrogen. Let the number densities of neutral atoms, ions, and electrons be $n_0$, $n_+$ and $n_e$ respectively. Also let $n_H$ be the number density of H nuclei. By charge neutrality

$$n_+ = n_e. \qquad (7.14)$$

Also

$$n_H = n_0 + n_+. \qquad (15)$$

To measure the degree of ionization, define

$$f = \frac{n_+}{n_H}, \qquad (7.16)$$

so that $f$ ranges from 0 for completely neutral hydrogen to 1 for fully ionized hydrogen.

The Saha equation gives

$$\frac{n_+ n_e}{n_0} = 2\frac{g_+}{g_0}\left(\frac{2\pi m_e kT}{h^2}\right)^{3/2} \exp\left(-\frac{\chi_H}{kT}\right). \qquad (7.17)$$

If we fix the spin of the proton, then $g_+ = 1$ and $g_0 = 2$. (Alternatively, if we do not fix the proton spin but allow it two spin states, then $g_+ = 2$ and $g_0 = 4$.) Hence

$$\frac{f^2}{1-f} = \frac{1}{n_H}\left(\frac{2\pi m_e kT}{h^2}\right)^{3/2} \exp\left(-\frac{\chi_H}{kT}\right) = 4.0 \times 10^{-9}\frac{T^{3/2}}{\rho} \exp\left(-\frac{1.610^5}{T}\right), \qquad (7.18)$$

where $T$ is in units of K and the density is in units of g cm$^{-3}$.

To locate the ionization zone on the log $\rho$–log $T$ diagram, consider the curves for $f = 0.1$ and $f = 0.9$. These are shown in figure 7.1 together with a solar model. The cross marks the location of the solar photosphere. We see that hydrogen is mainly neutral in the photosphere and the ionization zone ranges in temperature from about 11 000–51 000 K.

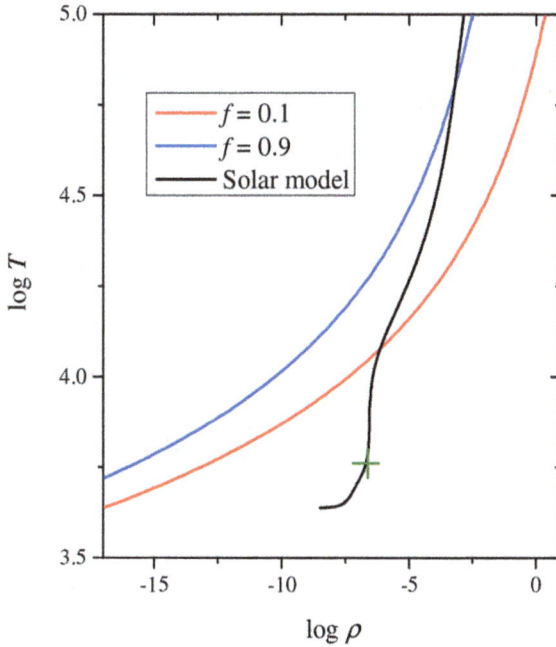

**Figure 7.1.** Boundaries of the H ionization zone. Also shown is the locus of a solar model.

## 7.6 The effect of ionization on the adiabatic gradient

Here we consider the effects of ionization on the equation of state and in particular the adiabatic gradient, which enters into the Schwarzschild criterion for convective instability. The gas pressure is

$$p = (n_H + n_e)kT = (1 + f)n_H kT = (1 + f)\frac{k}{m_H}\rho T. \tag{7.19}$$

We see immediately that the mean molecular weight depends on the degree of ionization.

This expression for the pressure can be used to find the density differential

$$\frac{d\rho}{\rho} = \frac{dp}{p} - \frac{f}{(1+f)}\frac{df}{f} - \frac{dT}{T}. \tag{7.20}$$

The internal energy per unit volume is

$$\varepsilon = \frac{3}{2}(n_H + n_e)kT - n_0\chi_H. \tag{7.21}$$

The last term on the right-hand side is the potential energy of the neutral atoms. Here the zero energy level has been taken to be at the bottom of the continuum.

The internal energy per unit mass is then

$$u = \frac{3}{2}(1+f)\frac{kT}{m_H} - (1-f)\frac{\chi_H}{m_H}. \tag{7.22}$$

For an adiabatic change

$$du = \frac{p}{\rho^2}d\rho. \tag{7.23}$$

Hence

$$\frac{3}{2}(1+f)\frac{kT}{m_H}\frac{dT}{T} + \left(\frac{3}{2}\frac{kT}{m_H} + \frac{\chi_H}{m_H}\right)df = (1+f)\frac{kT}{m_H}\frac{d\rho}{\rho}$$

$$= (1+f)\frac{kT}{m_H}\frac{dp}{p} - \frac{kT}{m_H}df - (1+f)\frac{kT}{m_H}\frac{dT}{T}. \tag{7.24}$$

Collecting like terms together, we obtain

$$\frac{5}{2}\frac{dT}{T} + \left(\frac{5}{2} + \frac{\chi_H}{kT}\right)\frac{df}{1+f} = \frac{dp}{p}. \tag{7.25}$$

Using the expression for the pressure (equation (7.19)) to eliminate the density from the Saha equation gives

$$\frac{f^2}{1-f^2} = \frac{kT}{p}\left(\frac{2\pi m_e kT}{h^2}\right)^{3/2}\exp\left(-\frac{\chi_H}{kT}\right). \tag{7.26}$$

Taking the natural log of both sides gives

$$2\ln f - \ln\left(1-f^2\right) = \frac{5}{2}\ln T - \ln p - \frac{\chi_H}{kT} + C, \tag{7.27}$$

where $C$ is a constant. In differential form, this is

$$\frac{2}{f(1-f)(1+f)}df = \left(\frac{5}{2} + \frac{\chi_H}{kT}\right)\frac{dT}{T} - \frac{dp}{p}. \tag{7.28}$$

On eliminating $df$ from equations (7.25) and (7.28), we find for an adiabatic change

$$\left[5 + f(1-f)\left(\frac{5}{2} + \frac{\chi_H}{kT}\right)^2\right]\frac{dT}{T} = \left[2 + f(1-f)\left(\frac{5}{2} + \frac{\chi_H}{kT}\right)\right]\frac{dp}{p}. \tag{7.29}$$

Hence

$$\nabla_{ad} = \left.\frac{\partial \ln T}{\partial \ln p}\right|_s = \frac{2 + f(1-f)\left(\frac{5}{2} + \frac{\chi_H}{kT}\right)}{5 + f(1-f)\left(\frac{5}{2} + \frac{\chi_H}{kT}\right)^2}. \tag{7.30}$$

We see that in completely neutral material ($f = 0$) or in fully ionized material ($f = 1$), we recover the value for an ideal gas, $\nabla_{ad} = 2/5$. At the onset of ionization, which we take to be when $f = 0.1$, $kT \sim 0.07\chi_H$ and $\nabla_{ad} \sim 0.1$. We conclude that the adiabatic gradient can be much reduced in ionization zones, which makes convection more likely. (We shall see later that the opacity is increased in ionization zones, which also makes convection more likely.)

## 7.7 The effect of ionization on the specific heat

The specific heat at constant pressure is

$$C_p = T \left.\frac{\partial s}{\partial T}\right|_p = \left.\frac{\partial u}{\partial T}\right|_p - \frac{p}{\rho^2} \left.\frac{\partial \rho}{\partial T}\right|_p. \tag{7.31}$$

By similar manipulations to those used to derive the adiabatic gradient, we obtain

$$C_p = \frac{5}{2}\frac{k}{m_u}(1 + f)\left[1 + \frac{1}{5}f(1 - f)\left(\frac{5}{2} + \frac{\chi_H}{kT}\right)^2\right]. \tag{7.32}$$

For completely neutral hydrogen $C_p = 5k/2m_u$. The specific heat at constant pressure is twice this for completely ionized hydrogen, because there are twice as many free particles per unit mass. At the onset of ionization, $C_p \sim 7(5k/2m_u)$. Hence ionization greatly increases the specific heat. This is because added heat goes mainly into lifting electrons out of the potential well of the atomic nucleus rather than increasing the kinetic energy of the atoms.

## 7.8 Pressure ionization

At the center of the Sun, $T \approx 1.5 \times 10^7$ K and $\rho \approx 10^2$ g cm$^{-3}$. The Saha equation for hydrogen gives $f = 0.75$, i.e. 25% of the hydrogen atoms are neutral even though the temperature is 100 times that corresponding to the ionization potential. Is this reasonable? Let us compare the size of an atom with the average distance between nuclei at the solar center. The atomic radius (according to a simple Bohr model) is $r_H = 0.5$ Å $= 5 \times 10^{-9}$ cm. The volume per proton is $v = m_H/\rho \approx 10^{-26}$ cm$^3$, which corresponds to a radius of $1.4 \times 10^{-9}$ cm. This is much less than the size of a hydrogen atom and hence electron orbitals from neighboring atoms would overlap and we cannot then tell which nuclei electrons belong to. The electrons are essentially free. This effect is called *pressure ionization* and is similar to what happens in a metal.

## 7.9 Free energy approach to ionization

We can derive the Saha equation for hydrogen by considering the Helmholtz free energy, $F$, of a number of hydrogen atoms, nuclei and electrons in a fixed volume $V$.

For $N$ point particles the internal energy and entropy are given by

$$U = \frac{3}{2}NkT, \tag{7.33}$$

and

$$S = Nk \ln\left(c\frac{V}{N}T^{3/2}\right), \tag{7.34}$$

where $c$ is a constant (see section 6.2).

Let $N_0$, $N_+$, $N_e$ be the number of atom nuclei and electrons in $V$, respectively. The free energy due the thermal motions of the particles is

$$F_1 = \frac{3}{2}(N_0 + N_+ + N_e)kT - N_0kT \ln\left(c_0\frac{V}{N_0}T^{3/2}\right) - N_+kT \ln\left(c_+\frac{V}{N_+}T^{3/2}\right)$$

$$- N_ekT \ln\left(c_e\frac{V}{N_e}T^{3/2}\right). \tag{7.35}$$

To obtain the total free energy, we have to add the contribution, $F_2$, from the bound states of the atoms.

Making the approximation that all the atoms are in their ground states,

$$F_2 = -N_0\chi_H. \tag{7.36}$$

(If $U$ is independent of $T$, the entropy is zero or a function of $V$ alone, so that $F = U$.)

In equilibrium $F$ is a minimum subject to the constraints of charge neutrality and conservation of nuclei. These constraints are

$$N_+ = N_e,$$
$$N_0 + N_+ = N_H, \tag{7.37}$$

where $N_H$ is the total number of hydrogen nuclei, whether bound or free. The easiest way to take these constraints into account is to use the degree of ionization

$$f = \frac{N_+}{N_H}. \tag{7.38}$$

The total free energy is then given by

$$\frac{F}{N_HkT} = \frac{3}{2}(1 + f) - (1 - f)\ln\left(c_0\frac{V}{(1-f)N_H}T^{3/2}\right) - f \ln\left(c_+\frac{V}{fN_H}T^{3/2}\right)$$

$$- f \ln\left(c_e\frac{V}{fN_H}T^{3/2}\right) - (1 - f)\frac{\chi_H}{kT}$$

$$= \frac{3}{2}(1 + f) - (1 - f)\ln\left(c_0\frac{V}{N_H}T^{3/2}\right) - f \ln\left(c_+\frac{V}{N_H}T^{3/2}\right) - f \ln\left(c_e\frac{V}{N_H}T^{3/2}\right)$$

$$+ (1 - f)\ln(1 - f) + 2f \ln f - (1 - f)\frac{\chi_H}{kT}. \tag{7.39}$$

To find the minimum value of the free energy we need to set its derivative with respect to $f$ to zero:

$$\frac{\partial}{\partial f}\left(\frac{F}{N_HkT}\right) = \frac{3}{2} + \ln\left(c_0\frac{V}{N_H}T^{3/2}\right) - \ln\left(c_+\frac{V}{N_H}T^{3/2}\right) - \ln\left(c_e\frac{V}{N_H}T^{3/2}\right)$$

$$- \ln(1 - f) + 2 \ln f + 1 + \frac{\chi_H}{kT}$$

$$= 0. \tag{7.40}$$

From this, we obtain

$$\frac{f^2}{1-f} = \frac{c_+ c_e}{c_0} \frac{1}{n_H} T^{3/2} \exp\left(-\frac{5}{2} - \frac{\chi_H}{kT}\right), \qquad (7.41)$$

which is the same form as the earlier expression provided the constants $c_0$, $c_+$, $c_e$ are given appropriate values. Here $n_H = N_H/V$ is the number density of hydrogen nuclei, bound and free.

## 7.10 A crude model for inclusion of pressure ionization in a thermodynamically consistent way[2]

The van der Waals equation of state is

$$p = \frac{NkT}{V - Nb} - a\left(\frac{N}{V}\right)^2, \qquad (7.42)$$

where $b$ is the volume occupied by a particle and $a$ is constant. This equation of state is formulated by assuming that the particles in a gas behave like hard spheres that weakly attract each other.

To take into account just the excluded volume effects, we let $a = 0$. The pressure is related to the free energy by

$$p = -\left.\frac{\partial F}{\partial V}\right|_T. \qquad (7.43)$$

Integration of the pressure with respect to $V$, keeping $T$ fixed, gives

$$F = -NkT \ln(V - Nb) + \Im(T, N), \qquad (7.44)$$

where $\Im$ is not dependent on $V$. Comparison with the ideal gas free energy shows that the first term in the van der Waals equation of state is obtained by replacing $V$ by $V - Nb$.

Making this replacement for the hydrogen atoms (the only extended species of particles under consideration), we have that our model for the free energy of our mixture of atoms, nuclei, and electrons is

$$\frac{F}{N_H kT} = \frac{3}{2}(1 + f) - (1 - f)\ln\left(c_0 \frac{V - (1 - f)N_H v_H}{N_H} T^{3/2}\right) - f \ln\left(c_+ \frac{V}{N_H} T^{3/2}\right)$$

$$- f \ln\left(c_e \frac{V}{N_H} T^{3/2}\right) + (1 - f)\ln(1 - f) + 2f \ln f - (1 - f)\frac{\chi_H}{kT}, \qquad (7.45)$$

where $v_H$ is the volume of a hydrogen atom.

---

[2] For a realistic treatment of pressure dissociation and ionization in hydrogen see [3, 4].

Now we find that the condition for thermodynamic equilibrium is

$$\frac{\partial}{\partial f}\left(\frac{F}{N_H kT}\right) = \frac{5}{2} + \ln\left(c_0 \frac{V - (1-f)N_H \nu_H}{N_H} T^{3/2}\right) - \frac{N_H \nu_H (1-f)}{V - (1-f)N_H \nu_H}$$

$$- \ln\left(c_+ \frac{V}{N_H} T^{3/2}\right) - \ln\left(c_e \frac{V}{N_H} T^{3/2}\right) - \ln(1-f) + 2\ln f + \frac{\chi_H}{kT}$$

$$= 0, \tag{7.46}$$

which leads to

$$\frac{f^2}{1-f} = \frac{1}{1-(1-f)n_H \nu_H} \exp\left[\frac{n_H \nu_H (1-f)}{1-(1-f)n_H \nu_H}\right] \cdot \frac{c_+ c_e}{c_0} \frac{1}{n_H} T^{3/2} \exp\left(-\frac{5}{2} - \frac{\chi_H}{kT}\right). \tag{7.47}$$

A comparison with the Saha equation indicates that the ground state occupation probability is

$$w_0 = \left[1 - (1-f)n_H \nu_H\right] \exp\left[-\frac{(1-f)n_H \nu_H}{1-(1-f)n_H \nu_H}\right] = \left[1 - n_0 \nu_H\right] \exp\left[-\frac{n_0 \nu_H}{1 - n_0 \nu_H}\right]. \tag{7.48}$$

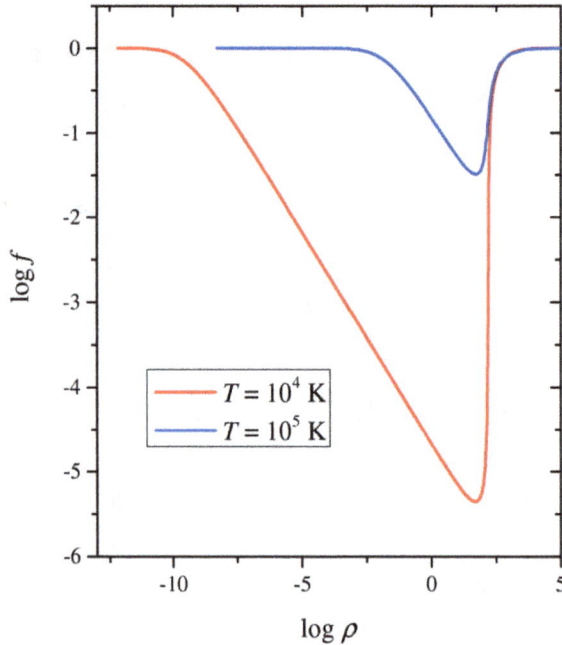

**Figure 7.2.** Degree of ionization of hydrogen from a simple model for pressure ionization.

We see that this goes very rapidly to zero as $n_0 v_H \to 1$. This sets an upper limit on the number density of hydrogen atoms. Figure 7.2 shows the degree of ionization plotted against density for temperatures $10^4$ and $10^5$ K. We see that at low enough density, hydrogen is completely ionized. As density increases, the degree of ionization decreases due to increasing recombination. This continues until the atoms begin to overlap. At this point pressure ionization occurs and there is a rapid increase in the degree of ionization.

## Bibliography

[1] López-Ruiza R, Sañudob J and Calbet X 2008 *Am. J. Phys.* **76** 8
[2] Hummer D G and Mihalas D 1988 *Astrophys. J.* **331** 794
[3] Saumon D and Chabrier G 1991 *Phys. Rev. A* **44** 5122
[4] Saumon D and Chabrier G 1992 *Phys. Rev. A* **46** 2084

# Chapter 8

## The degenerate electron gas

### 8.1 Introduction

In dense material such as is found in WDs and the cores of red giant stars, the electrons are sufficiently close together that the quantum nature of phase space must be taken into account. This leads to what is called *degenerate electron pressure*. Since electrons are spin ½ particles they obey Fermi–Dirac statistics and Pauli's exclusion principle. For free electrons, Pauli's exclusion principle gives that no more than two electrons can occupy a volume of $h^3$ in phase space. Here $h$ is Planck's constant.

To visualize what this means, restrict to one dimension in momentum space and consider a unit volume of position space. Figure 8.1 shows the distribution in momentum for a fixed non-zero temperature and four values of electron density.

At low density, the distribution is Maxwellian. The Pauli exclusion principle sets a ceiling on density in momentum space. At high density all states with momentum less than a threshold are occupied and very few electrons have momentum greater than this threshold.

We say that the electrons are *degenerate* if the Pauli exclusion principle significantly modifies the momentum distribution from Maxwellian.

### 8.2 Complete electron degeneracy

*Complete electron degeneracy* occurs at $T = 0$. In this case all states with momentum less than the threshold are filled and states with momentum greater than the threshold are empty. The threshold momentum, $q_f$, is called the *Fermi momentum*.

The electron number density for complete degeneracy is

$$n_e = \frac{2}{h^3} \int_0^{q_f} 4\pi q^2 \mathrm{d}q = \frac{2}{h^3} \frac{4\pi}{3} q_f^3. \tag{8.1}$$

doi:10.1088/978-1-6817-4105-5ch8

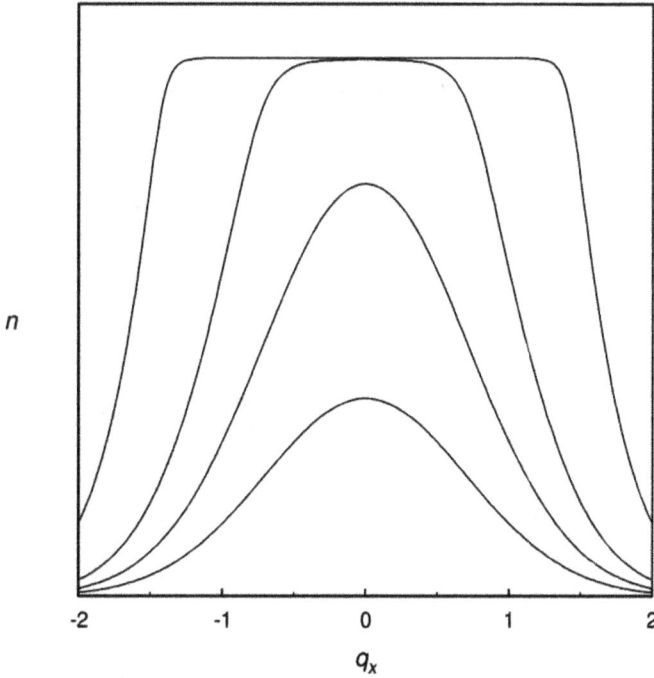

**Figure 8.1.** Schematic showing the effects of degeneracy on the electron momentum distribution.

In a similar way to that used to find the radiation pressure, to calculate the electron pressure we consider the momentum flux in the $x$-direction

$$P_e = \iiint v_x q_x \frac{2}{h^3} dq_x dq_y dq_z. \tag{8.2}$$

By isotropy

$$3P_e = \iiint \left( v_x q_x + v_y q_y + v_z q_z \right) \frac{2}{h^3} dq_x dq_y dq_z = \int_0^{q_f} v q \frac{2}{h^3} 4\pi q^2 dq. \tag{8.3}$$

The electrons might be relativistic and hence we must include the Lorentz factor in relating velocity and momentum

$$q = mv \left( 1 - \frac{v^2}{c^2} \right)^{-1/2}, \tag{8.4}$$

$$\frac{v}{c} = \frac{q}{mc} \left[ 1 + \left( \frac{q}{mc} \right)^2 \right]^{-1/2}, \tag{8.5}$$

where $m$ is the electron mass.

Hence for completely degenerate electrons,

$$p_e = \frac{8\pi}{3h^3} \int_0^{q_f} vq^3 dq = \frac{8\pi}{3mh^3} \int_0^{q_f} q^4 \left[1 + \left(\frac{q}{mc}\right)^2\right]^{-1/2} dq. \tag{8.6}$$

The integral is evaluated by making the substitution $q = mc \sinh \theta$, which gives

$$p_e = \frac{\pi m^4 c^5}{3h^3} f(x), \tag{8.7}$$

where

$$f(x) = x(2x^2 - 3)(x^2 + 1)^{1/2} + 3 \sinh^{-1}x, \tag{8.8}$$

and

$$x = \frac{q_f}{mc}, \tag{8.9}$$

is a dimensionless Fermi momentum or 'relativity parameter'.

The electron number density is given in terms of the matter density by

$$n_e = \frac{\rho}{m_u \mu_e}, \tag{8.10}$$

where $\mu_e$ is the mean molecular weight per electron. This allows us to write $q_f$ and $x$ in terms of the matter density. We find

$$x = \left(\frac{3}{8\pi}\right)^{1/3} \frac{h}{mc} \left(\frac{\rho}{m_u \mu_e}\right)^{1/3}. \tag{8.11}$$

In cgs units,

$$p_e = 6.00 \times 10^{22} f(x) \text{ dyne cm}^{-2}, \tag{8.12}$$

and

$$x = 1.01 \times 10^{-2} \left(\frac{\rho}{\mu_e}\right)^{1/3}. \tag{8.13}$$

The energy density (energy per unit volume) of the completely degenerate electrons (including their rest mass energy) is

$$\varepsilon = \frac{2}{h^3} \int_0^{q_f} mc^2 \left(1 + \frac{q^2}{m^2 c^2}\right)^{1/2} 4\pi q^2 dq. \tag{8.14}$$

Making the same substitution as for the pressure, we obtain

$$\varepsilon = \pi \left(\frac{mc}{h}\right)^3 mc^2 \left[x(1 + x^2)^{1/2}(1 + 2x^2) - \sinh^{-1}x - \frac{8}{3}x^3\right] + \pi \left(\frac{mc}{h}\right)^3 mc^2 \frac{8}{3}x^3, \tag{8.15}$$

where the second term of the right-hand side is the rest mass energy density.

## 8.3 Limiting forms

The limits $x \ll 1$ and $x \gg 1$ correspond to non-relativistic and relativistic electrons, respectively. The relation between pressure and density in these limits is most easily obtained by considering equation (8.6) in these limits.

For non-relativistic electrons, $q \ll mc$, and so

$$p_e \approx \frac{8\pi}{3mh^3} \int_0^{q_f} q^4 dq = \frac{8\pi}{15mh^3} q_f^5 = \frac{8\pi m^4 c^5}{15h^3} x^5 = \frac{h^2}{5m}\left(\frac{3}{8\pi}\right)^{2/3}\left(\frac{\rho}{m_u \mu_e}\right)^{5/3}$$

$$= 1.00 \times 10^{13} \left(\frac{\rho}{\mu_e}\right)^{5/3} \text{ dyne cm}^{-2}. \tag{8.16}$$

For highly relativistic electrons,

$$p_e = \frac{8\pi c}{3h^3} \int_0^{q_f} q^3 dq = \frac{2\pi c}{3h^3} q_f^4 = \frac{2\pi m^4 c^5}{3h^3} x^4 = \frac{1}{4}\left(\frac{3}{8\pi}\right)^{1/3} hc\left(\frac{\rho}{m_u \mu_e}\right)^{4/3}$$

$$= 1.24 \times 10^{15} \left(\frac{\rho}{\mu_e}\right)^{4/3} \text{ dyne cm}^{-2}. \tag{8.17}$$

We can estimate the density, $\rho_{nr\_r}$, at which the transition from non-relativistic to relativistic electrons occurs by comparing these two expressions for the electron pressure. We find

$$\frac{\rho_{nr\_r}}{\mu_e} \approx 1.9 \times 10^6 \text{ g cm}^{-3}. \tag{8.18}$$

The limiting forms of the internal energy per unit volume are

$$\varepsilon = \frac{8\pi m^3 c^3}{h^3} mc^2 \left(\frac{x^3}{3} + \frac{x^5}{10} + \cdots\right) \tag{8.19}$$

for $x \ll 1$ and

$$\varepsilon = \frac{2\pi m^3 c^3}{h^3} mc^2 (x^4 + x^2 + \cdots), \tag{8.20}$$

for $x \gg 1$. Note that expressions (8.19) and (8.20) include the rest mass energy density.

## 8.4 The contribution from nuclei at zero temperature

The contribution to the pressure from completely degenerate nuclei is negligible. To see why, suppose for simplicity that the nuclei are protons of mass $M$. By charge neutrality, the proton number density is the same as the electron number density and hence the Fermi momentum is the same. The dimensional Fermi parameter for the protons is smaller than that for the electrons by a factor $m/M$. Hence the nuclei are much less relativistic than the electrons. From equation (8.16), the pressure from the nuclei is

$$P_{\text{nuc}} \approx \frac{8\pi}{15Mh^3}q_0^5 = \frac{m}{M}\frac{8\pi m^4 c^5}{15h^3}x^5 = 5.51 \times 10^9 \left(\frac{\rho}{\mu_e}\right)^{5/3} \text{dyne cm}^{-2}. \quad (8.21)$$

This is small compared to the pressure from non-relativistic electrons and becomes comparable to that from relativistic electrons when

$$\frac{\rho}{\mu_e} \approx 10^{16} \text{ g cm}^{-3}.$$

This is greater than the mean densities of atomic nuclei and hence for cold WDs, the contribution to the pressure from completely degenerate nuclei is negligible.

## 8.5 Transition from non-degeneracy to degeneracy

At finite temperature and low enough density, the electron momentum distribution will be Maxwellian and the electron pressure can be found from the ideal gas law

$$P_e = n_e kT = \frac{\Re \rho T}{\mu_e}. \quad (8.22)$$

We can find the density, $\rho_{\text{nd\_d}}$, at which the electrons start to become degenerate by equating the expressions for electron pressure in equations (8.16) and (8.22):

$$\frac{h^2}{5m}\left(\frac{3}{8\pi}\right)^{2/3}\left(\frac{\rho_{\text{nd\_d}}}{m_u\mu_e}\right)^{5/3} = \frac{\Re \rho_{\text{nd\_d}} T}{\mu_e}. \quad (8.23)$$

This gives

$$\frac{\rho_{\text{nd\_d}}}{\mu_e} = \frac{8\pi m_u}{3}\left(5\frac{mkT}{h^2}\right)^{3/2} = 2.5 \times 10^{-8} T^{3/2} \text{ g cm}^{-3}. \quad (8.24)$$

At the solar center, $T \approx 1.5 \times 10^7$ K, and so $\rho_{\text{nd\_d}} \sim 2 \times 10^3$ g cm$^{-3}$, which is greater than the solar central density by a factor of about 20. Hence the electrons in the Sun are non-degenerate.

## 8.6 Effects of degeneracy on the adiabatic gradient and the first adiabatic exponent

To see how electron degeneracy affects the adiabatic gradient and the first adiabatic exponent, first consider a situation in which the electrons are non-relativistic but sufficiently degenerate that equation (8.16) gives a good approximation to the electron pressure. Adding the contribution from the non-degenerate ions, the total pressure is

$$p = \frac{h^2}{5m}\left(\frac{3}{8\pi}\right)^{2/3}\left(\frac{\rho}{m_u\mu_e}\right)^{5/3} + \frac{\rho}{m_u\mu_{\text{ion}}}kT, \quad (8.25)$$

where $\mu_{\text{ion}}$ is the mean molecular weight per ion. The molecular weights are related by

$$\sum_k \frac{X_k}{A_k}(1 + Z_k) = \frac{1}{\mu} = \frac{1}{\mu_{\text{ion}}} + \frac{1}{\mu_e} = \sum_k \frac{X_k}{A_k} + \sum_k \frac{X_k Z_k}{A_k}, \qquad (8.26)$$

where $Z_k$ is the number of electrons freed from atoms of species $k$.

The internal energy per unit mass (excluding rest mass energy) is

$$u = \frac{3}{2} \frac{h^2}{5m\rho} \left(\frac{3}{8\pi}\right)^{2/3} \left(\frac{\rho}{m_u \mu_e}\right)^{5/3} + \frac{3}{2} \frac{kT}{m_u \mu_{\text{ion}}}. \qquad (8.27)$$

Comparing equations (8.25) and (8.27), we see that

$$u = \frac{3}{2} \frac{p}{\rho}, \qquad (8.28)$$

just as for the ideal gas. Since for an adiabatic change

$$du = \frac{p}{\rho} \frac{d\rho}{\rho}, \qquad (8.29)$$

we have

$$du = \frac{3}{2} \frac{dp}{\rho} - \frac{3}{2} \frac{p}{\rho} \frac{d\rho}{\rho} = \frac{p}{\rho} \frac{d\rho}{\rho}. \qquad (8.30)$$

We immediately see that

$$\Gamma_1 = \frac{5}{3}. \qquad (8.31)$$

Also, on using the expressions for $u$ and $p$ in equation (8.29), we obtain

$$\frac{h^2}{5m\rho}\left(\frac{3}{8\pi}\right)^{2/3}\left(\frac{\rho}{m_u \mu_e}\right)^{5/3} \frac{d\rho}{\rho} + \frac{3}{2} \frac{k}{m_u \mu_{\text{ion}}} dT$$

$$= \frac{h^2}{5m\rho}\left(\frac{3}{8\pi}\right)^{2/3}\left(\frac{\rho}{m_u \mu_e}\right)^{5/3} \frac{d\rho}{\rho} + \frac{k}{m_u \mu_{\text{ion}}} T \frac{d\rho}{\rho}. \qquad (8.32)$$

The first terms on each side cancel and we obtain

$$\frac{3}{2} \frac{dT}{T} = \frac{d\rho}{\rho}. \qquad (8.33)$$

Since

$$\nabla_{\text{ad}} = \left.\frac{\partial \ln T}{\partial \ln p}\right|_s = \left.\frac{\partial \ln T}{\partial \ln \rho}\right|_s \bigg/ \left.\frac{\partial \ln p}{\partial \ln \rho}\right|_s = \frac{1}{\Gamma_1} \left.\frac{\partial \ln T}{\partial \ln \rho}\right|_s, \qquad (8.34)$$

we find that $\nabla_{\text{ad}} = 2/5$, again the same as for an ideal gas.

Now consider a situation in which the electrons are highly relativistically degenerate. The total pressure and internal energy per unit mass are

$$p = \frac{1}{4}\left(\frac{3}{8\pi}\right)^{1/3} hc \left(\frac{\rho}{m_u \mu_e}\right)^{4/3} + \frac{\rho}{m_u \mu_{ion}} kT, \tag{8.35}$$

and

$$u = \frac{3}{4}\left(\frac{3}{8\pi}\right)^{1/3} \frac{ch}{\rho} \left(\frac{\rho}{m_u \mu_e}\right)^{4/3} + \frac{3}{2}\frac{kT}{m_u \mu_{ion}}. \tag{8.36}$$

For an adiabatic change, we obtain

$$\frac{1}{3}\frac{3}{4}\left(\frac{3}{8\pi}\right)^{1/3} \frac{ch}{\rho} \left(\frac{\rho}{m_u \mu_e}\right)^{4/3} \frac{d\rho}{\rho} + \frac{3}{2}\frac{kT}{m_u \mu_{ion}} \frac{dT}{T}$$
$$= \frac{1}{4}\left(\frac{3}{8\pi}\right)^{1/3} \frac{hc}{\rho} \left(\frac{\rho}{m_u \mu_e}\right)^{4/3} \frac{d\rho}{\rho} + \frac{kT}{m_u \mu_{ion}} \frac{d\rho}{\rho}. \tag{8.37}$$

Again, the first terms on each side cancel, and so

$$\frac{3}{2}\frac{dT}{T} = \frac{d\rho}{\rho}. \tag{8.33}$$

From equation (8.35),

$$dp = \frac{1}{3}\left(\frac{3}{8\pi}\right)^{1/3} hc \left(\frac{\rho}{m_u \mu_e}\right)^{4/3} \frac{d\rho}{\rho} + \frac{\rho kT}{m_u \mu_{ion}} \frac{d\rho}{\rho} + \frac{\rho kT}{m_u \mu_{ion}} \frac{dT}{T}, \tag{8.39}$$

so that for an adiabatic change

$$\frac{dp}{p} = \frac{\frac{1}{2}\left(\frac{3}{8\pi}\right)^{1/3} hc \left(\frac{\rho}{m_u \mu_e}\right)^{4/3} + \frac{5}{2}\frac{\rho kT}{m_u \mu_{ion}}}{\frac{1}{4}\left(\frac{3}{8\pi}\right)^{1/3} hc \left(\frac{\rho}{m_u \mu_e}\right)^{4/3} + \frac{\rho kT}{m_u \mu_{ion}}} \frac{dT}{T} = \frac{2p_e + \frac{5}{2}p_{ion}}{p_e + p_{ion}} \frac{dT}{T}. \tag{8.40}$$

Hence for highly relativistic degenerate electrons,

$$\nabla_{ad} = \frac{p_e + p_{ion}}{2p_e + \frac{5}{2}p_{ion}}, \tag{8.41}$$

and

$$\Gamma_1 = \frac{4p_e + 5p_{ion}}{3p_e + 3p_{ion}}. \tag{8.42}$$

If the electron pressure dominates (which is the usual case when the electrons are relativistically degenerate), then

$$\nabla_{ad} = \frac{1}{2}, \tag{8.43}$$

and

$$\Gamma_1 = \frac{4}{3}. \tag{8.44}$$

# Chapter 9

## Polytropes and the Chandrasekhar mass

### 9.1 Introduction

A polytropic equation of state is one in which the pressure has a power law dependence on density:

$$p = K\rho^{\gamma}, \tag{9.1}$$

where $\gamma$ and $K$ are constants. Examples are the two limiting forms for completely degenerate electrons and also the Eddington standard model in which $\beta = p_{gas}/p$ is uniform throughout the star.

### 9.2 The Lane–Emden equation

Since the pressure is a function of density alone, the structure of a polytrope can be found by solving the continuity and hydrostatic equilibrium equations:

$$\frac{dm}{dr} = 4\pi r^2 \rho, \tag{9.2}$$

and

$$\frac{dp}{dr} = -\frac{Gm\rho}{r^2}. \tag{9.3}$$

By re-arranging the hydrostatic equilibrium equation (9.3) to give an expression for $m$ and then using the polytropic equation of state, we obtain

$$m = -\frac{r^2}{G\rho}\frac{dp}{dr} = -\frac{\gamma K}{(\gamma - 1)G}r^2\frac{d\rho^{\gamma-1}}{dr}. \tag{9.4}$$

On eliminating $m$ from equation (9.2), we obtain

$$\frac{d}{dr}\left(r^2 \frac{d\rho^{\gamma-1}}{dr}\right) = -4\pi \frac{(\gamma-1)G}{\gamma K} r^2 \rho. \tag{9.5}$$

This equation is reduced to a dimensionless form by introducing the *polytropic index*

$$n = \frac{1}{\gamma - 1}, \tag{9.6}$$

and a length scale

$$a = \left[\frac{(n+1)K\rho_c^{\frac{1}{n}-1}}{4\pi G}\right]^{1/2}, \tag{9.7}$$

where $\rho_c$ is the central density.

By making the substitutions

$$\rho = \rho_c \theta^n, \tag{9.8}$$

and

$$r = a\xi, \tag{9.9}$$

equation (9.5) becomes

$$\frac{1}{\xi^2} \frac{d}{d\xi}\left(\xi^2 \frac{d\theta}{d\xi}\right) + \theta^n = 0. \tag{9.10}$$

This is the *Lane–Emden equation* for the structure of a polytrope of index $n$. The boundary conditions at the center are

$$\theta(0) = 1, \tag{9.11}$$

and

$$\frac{d\theta}{d\xi}(0) = 0. \tag{9.12}$$

The second condition comes from noting that $dP/dr = 0$ at the stellar center.

The solution to equation (9.10) depends only on the polytropic index $n$. For $n < 5$, the solution decreases monotonically and becomes zero at a finite value $\xi = \xi_1$, which corresponds to the surface of the star, where $\rho = 0$.

The radius of the star is

$$R = a\xi_1 = \left[\frac{(n+1)K}{4\pi G}\right]^{1/2} \rho_c^{(1-n)/2n} \xi_1. \tag{9.13}$$

The mass of the star is from equation (9.4)

$$M = -a\frac{\gamma K \rho_c^{\gamma-1}}{(\gamma-1)G}\xi_1^2\theta'(\xi_1) = 4\pi\left[\frac{(n+1)K}{4\pi G}\right]^{3/2}\rho_c^{(3-n)/2n}\xi_1^2\,|\theta'(\xi_1)|. \qquad (9.14)$$

Eliminating the central density from equations (9.13) and (9.14) gives the mass radius relation

$$M = 4\pi R^{(3-n)/(1-n)}\left[\frac{(n+1)K}{4\pi G}\right]^{n/(n-1)}\xi_1^{(3-n)/(1-n)}\xi_1^2\,|\theta'(\xi_1)|. \qquad (9.15)$$

## 9.3 Application to white dwarf stars

For low density WDs,

$$\gamma = \frac{5}{3}, \quad n = \frac{3}{2}, \quad \xi_1 = 3.654, \quad \xi_1^2\,|\theta'(\xi_1)| = 2.714, \quad K = \frac{1.00\times10^{13}}{\mu_e^{5/3}}. \qquad (9.16)$$

Hence

$$R = 1.12\ 10^9\left(\frac{\rho_c}{10^6\ \text{g cm}^{-3}}\right)^{-1/6}\left(\frac{\mu_e}{2}\right)^{-5/6}\text{cm}, \qquad (9.17)$$

$$M = 0.496\left(\frac{\rho_c}{10^6\ \text{g cm}^{-3}}\right)^{1/2}\left(\frac{\mu_e}{2}\right)^{-5/2}M_\odot, \qquad (9.18)$$

$$M = 0.701\left(\frac{R}{10^9\ \text{cm}}\right)^{-3}\left(\frac{\mu_e}{2}\right)^{-5}M_\odot. \qquad (9.19)$$

Clearly the radius decreases with increasing mass.

Since the transition from non-relativistic to relativistic electrons occurs at a density

$$\rho_{nrr} \approx 1.9\times10^6\,\mu_e\ \text{g cm}^{-3}, \qquad (9.20)$$

we see from equation (9.18) that, for $\mu_e = 2$, relativistic effects are important for WDs with mass greater than about 1.0 $M_\odot$.

For high mass WDs in which the electrons are relativistically degenerate

$$\gamma = \frac{4}{3}, \quad n = 3, \quad \xi_1 = 6.897, \quad \xi_1^2\,|\theta'(\xi_1)| = 2.018, \quad K = \frac{1.24\times10^{15}}{\mu_e^{4/3}}. \qquad (9.21)$$

Hence now

$$R = 3.35\times10^9\left(\frac{\rho_c}{10^6\ \text{g cm}^{-3}}\right)^{-1/3}\left(\frac{\mu_e}{2}\right)^{-2/3}\text{cm}, \qquad (9.22)$$

and

$$M = 1.457\left(\frac{\mu_e}{2}\right)^{-2} M_\odot. \tag{9.23}$$

Note that $M$ is independent of $\rho_c$ and hence $R$. As $\rho_c \to \infty$, the electrons become more and more relativistic throughout the star and the mass asymptotically approaches the value given in equation (9.23) as $R \to 0$. This mass limit is called the *Chandrasekhar limit*. It is an upper limit on the mass of cold WDs and was first derived by Chandrasekhar [1].

## Bibliography

[1] Chandrasekhar S 1931 *Astrophys. J.* **74** 81

Structure and Evolution of Single Stars
An introduction
**James MacDonald**

# Chapter 10

## Opacity

## 10.1 Introduction

In radiative regions of a star, the diffusion approximation to radiative transfer relates the luminosity to the temperature gradient:

$$L = -\frac{16\pi acr^2 T^3}{3\kappa\rho}\frac{dT}{dr}. \tag{10.1}$$

In this expression, $\kappa$ is the mean opacity of the stellar material. It depends on the composition of the material as well as the density and temperature.

## 10.2 The Rosseland mean opacity

If matter and radiation are in thermal equilibrium, the distribution of the specific intensity of the radiation with frequency is given by the Planck function

$$B_\nu(T) = \frac{2h\nu^3}{c^2}\frac{1}{e^{h\nu/kT} - 1}, \tag{10.2}$$

where $B_\nu(T)d\nu$ is the energy in frequency interval $\nu$ to $\nu + d\nu$ that crosses unit area in unit time into unit solid angle.

In exact thermal equilibrium there is equal flow of radiation in all directions. Inside a star there is a net outflow flow of radiation and hence there must be deviation from exact thermal equilibrium. However in stellar interiors the deviations from the Planck distribution are small enough to be neglected.

Photons of different energy will be absorbed at different rates that depend on the thermal state of the stellar material. Hence opacity is a function of frequency, which we will denote by $\kappa_\nu$. In equation (10.1), the opacity is an average over frequency. Here we will sketch how this averaging should be done. There are two factors that need to be taken into consideration.

The first consideration is that we should average the 'conductivity', i.e. the inverse of the opacity instead of the opacity. This can be shown from a simple picture in which the opacity is high in a single frequency interval and low at all other frequencies. Photons of energies corresponding to the high opacity frequencies will be readily absorbed (or scattered) and photons of other energies are readily transmitted. In thermal equilibrium, the energy of the absorbed photons must be re-emitted (otherwise the temperature would change) and the emitted photons will have a Planck distribution. Hence the energy flux is mainly carried by photons with frequencies at which the opacity is low, i.e. frequencies at which the 'conductivity' is high. This indicates that rather than averaging the opacity, it is more appropriate to consider an average of its inverse so as to give greater weight to frequencies where the opacity is lower than average.

The second consideration is how the energy distribution of the photons enters into the average. As shown earlier, the radiative energy flux is proportional to the gradient of the frequency-integrated Planck function. Hence the flux carried by photons of frequency $\nu$ is such that

$$F_\nu \propto -\frac{dB_\nu(T)}{dr} = -\frac{dB_\nu}{dT}\frac{dT}{dr}. \tag{10.3}$$

These two considerations indicate that the appropriate opacity average is

$$\frac{1}{\kappa} = \frac{\displaystyle\int_0^\infty \frac{1}{\kappa_\nu}\frac{dB_\nu}{dT}d\nu}{\displaystyle\int_0^\infty \frac{dB_\nu}{dT}d\nu}. \tag{10.4}$$

This is the Rosseland mean opacity [1].

We see that the important frequency intervals for determining the opacity will be those where $\kappa_\nu$ is small and those where $dB_\nu/dT$ is large. From equation (9.2),

$$\frac{dB_\nu}{dT} \propto \frac{u^5 e^u}{(e^u - 1)^2} = g(u), \tag{10.5}$$

where $u = h\nu/kT$. The function $g(u)$ is maximum at $u = 4.928$. Hence frequencies for which $h\nu \approx 5kT$ are important for determining the opacity. For comparison, the Planck function peaks near $u = 2.82$.

## 10.3 Opacity mechanisms

Radiative opacity is due to both scattering and absorption of photons. Absorption processes include bound–bound, bound–free and free–free electron transitions.

In bound–bound absorption, a photon is destroyed in exciting a bound state of an atom or ion to a higher energy bound state. The energy of the photon is equal to the difference in the energy of the two bound states

$$h\nu_{bb} = E_2 - E_1. \tag{10.6}$$

These transitions are responsible for absorption lines in stellar spectra. (If the photon is re-emitted by a downward transition, the photon has essentially just been scattered. However the excited state can be collisionally de-excited in which the energy goes into the kinetic energy of the colliding particles. In this case, we have true absorption of the photon. The relative rates of collisional and photo-de-excitation depend on the particle density.)

In bound–free absorption, a photon is destroyed in removing an electron from an atom or an ion. This is also called photo-ionization. The photon energy must exceed a threshold for this process to occur and gives rise to absorption edges in stellar spectra.

In free–free absorption, a photon is destroyed in moving an electron to a higher continuum energy state in the vicinity of an ion. The presence of the ion is required to simultaneously conserve energy and momentum. This process cannot occur for unaccompanied electrons.

An electromagnetic wave incident on an electron accelerates the electron. Since accelerated charged particles radiate, radiation will be emitted. Since the energy of the emitted radiation comes from that of the incident electromagnetic wave, part of the incident wave has been scattered by the electron. If the energy of the incident photons is much less than the rest mass energy of the electron, the electron is barely moved by the collision and hence the collision cross section is independent of the photon frequency. This is the case for *Thomson scattering*.

For true absorption processes, stimulated emission reduces the opacity. Stimulated emission occurs when a photon induces a downward transition in energy. For a bound–bound transition, the stimulating photon must have the same energy as the emitted photon.

## 10.4 Electron scattering opacity

The Thomson cross section is

$$\sigma_e = \frac{8\pi}{3}\left(\frac{e^2}{m_e c^2}\right)^2 = 6.65 \times 10^{-25}\,\text{cm}^2. \tag{10.7}$$

The opacity is related to the cross section by

$$\kappa\rho = n\sigma. \tag{10.8}$$

Hence the electron scattering opacity is

$$\kappa_{es} = \frac{n_e}{\rho}\sigma_e \simeq 0.20(1 + X)\,\text{cm}^2\,\text{g}^{-1}. \tag{10.9}$$

## 10.5 Free–free opacity

For a mixture of electrons and charged particles of atomic mass $A$ and charge $Z$, the free–free opacity is

$$\kappa_\nu \approx 2 \times 10^{20}\frac{Z^2}{A}\frac{n_e}{T^{1/2}}\frac{1}{\nu^3}\left(1 - e^{-h\nu/kT}\right)\text{cm}^2\,\text{g}^{-1}, \tag{10.10}$$

where $n_e$ is the electron number density in units of electrons per $cm^3$.

If this were the only opacity process, the integrals in the Rosseland mean opacity can be evaluated to give

$$\kappa_{ff} \approx 7.5 \times 10^{22} \frac{Z^2}{A} \frac{\rho}{\mu_e} \frac{1}{T^{7/2}} \; cm^2 \, g^{-1}. \qquad (10.11)$$

Since this was first derived by Hendrik Kramers, a dependence of opacity on density and temperature of this form is called a Kramers' opacity law.

If we compare the free–free opacity to that from electron scattering for pure hydrogen

$$\kappa_{ff} \approx 7.5 \times 10^{22} \frac{\rho}{T^{7/2}} \; cm^2 \, g^{-1}, \qquad (10.12)$$

$$\kappa_{es} = 0.40 \; cm^2 \, g^{-1}, \qquad (10.13)$$

we see that the transition locus is

$$\rho = 20 \left( \frac{T}{10^7 \, K} \right)^{7/2} \; g \, cm^{-3}. \qquad (10.14)$$

This shows that both electron scattering and free–free absorption are important contributors to the opacity at the solar center. Bound–free and bound–bound absorption are less important because of the high ionization state of the solar interior.

## 10.6 Bound–free opacity

For a bound state of energy $E$ below the continuum, the photo-ionization cross section is of form

$$\sigma_{bf}(\nu) = \begin{cases} A\nu^{-n} & \text{for } h\nu \geqslant E, \\ 0 & \text{for } h\nu < E. \end{cases} \qquad (10.15)$$

Typically $n \approx 3$. The opacity also depends on the occupation number of the state which is given by a Boltzmann factor. The contribution to the opacity from a single ionic species is as shown schematically in figure 10.1.

Low energy photons can only ionize highly excited states. In the absence of interactions, these states can be well populated at high enough temperature and hence the opacity can be relatively high. As the photon energy increases, the cross section for photo-ionization decreases and hence the opacity decreases with frequency except when the photon energy first exceeds the threshold for ionization of a more tightly bound state. The schematic is based on the ionization of atomic hydrogen at low density and a temperature of 12 000 K. The frequency unit is the frequency of the Lyman limit. The details of the opacity at low frequency are relatively unimportant for the Rosseland mean opacity because the peak of the $dB_\nu/dT$ weighting factor is at much higher frequency.

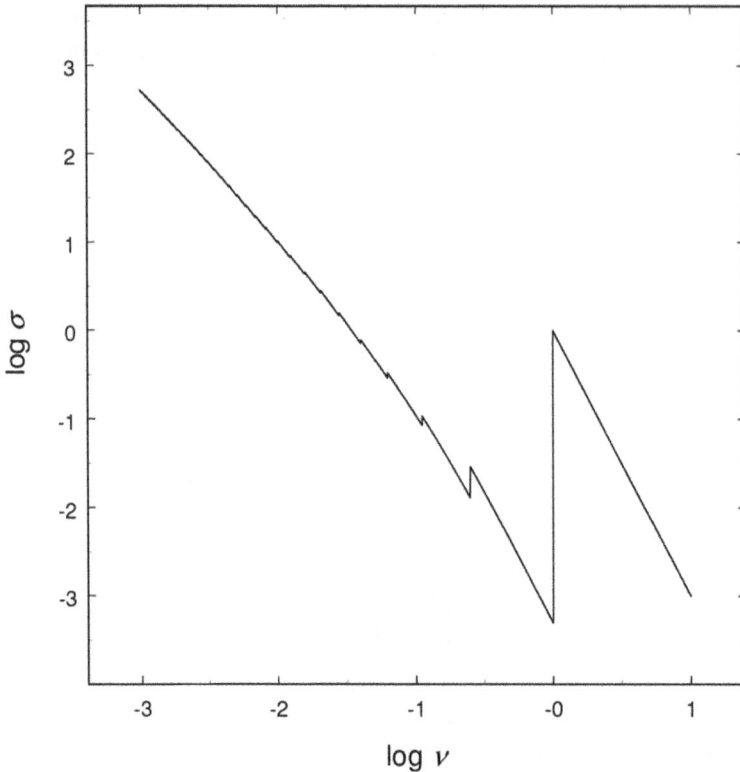

**Figure 10.1.** Schematic showing the variation of bound–free opacity with frequency.

Since many species contribute to the bound–free opacity, the calculation of its contribution to the Rosseland mean opacity is difficult and requires careful consideration of the effects of particle interactions on the populations of excited states as well as on ionization.

Since bound–free opacity is associated with photo-ionization, there is usually an increase in opacity for temperature–density conditions associated with ionization of a common element.

## 10.7 Bound–bound opacity

The opacity for a bound–bound transition can be estimated by considering the electron to be a damped harmonic oscillator driven by the oscillating electric field of the electromagnetic radiation. The equation of motion of the electron is

$$m_e\left(\ddot{x} + \omega_0^2 x\right) = eE_0 e^{i\omega t} - m_e\gamma\dot{x}, \qquad (10.16)$$

where $\omega_0$ is the natural frequency of the oscillator and $\omega$ is the frequency of the driving electric field of amplitude $E_0$. The damping constant $\gamma$ arises because, as it oscillates, the electron will radiate away energy. In the classical picture, $\gamma$ is estimated as follows.

According to classical electromagnetic theory, the total power radiated in all directions by a particle undergoing acceleration, $a$, is

$$P = \frac{2e^2a^2}{3c^3}. \tag{10.17}$$

For a particle undergoing harmonic motion with frequency $\omega$ and amplitude $x_0$, the acceleration is proportional to $x_0\omega^2$ and the time averaged power radiated is

$$\langle P \rangle = \frac{e^2x_0^2\omega^4}{3c^3}. \tag{10.18}$$

Equating this with the time averaged rate at which work is done by the damping force, we obtain

$$\gamma = \frac{2e^2\omega^2}{3m_ec^3}. \tag{10.19}$$

For spectral lines in the optical part of the spectrum, $\gamma \sim 10^8 \text{ s}^{-1}$, which is small compared to the frequency of the spectral line, $\omega \sim 3 \times 10^{15} \text{ s}^{-1}$.

The (long term) physical solution to equation (10.16) is

$$x = \text{Re}\left(\frac{e}{m_e}\frac{E_0e^{i\omega t}}{\omega^2 - \omega_0^2 + i\gamma\omega}\right). \tag{10.20}$$

Since the particle acceleration is $a = -\omega^2x$, the time averaged power radiated is, from equation (10.17),

$$\langle P(\omega) \rangle = \frac{e^4\omega^4}{3m_e^2c^3}\frac{E_0^2}{\left(\omega^2 - \omega_0^2\right)^2 + \gamma^2\omega^2}. \tag{10.21}$$

This is to be interpreted as the power scattered out of a beam of energy density $E_0^2/8\pi$, so that

$$\langle P(\omega) \rangle = \frac{cE_0^2}{8\pi}\sigma(\omega). \tag{10.22}$$

Comparing equations (10.21) and (10.22), we find

$$\sigma(\omega) = \frac{8\pi e^4}{3m_e^2c^4}\frac{\omega^4}{\left(\omega^2 - \omega_0^2\right)^2 + \gamma^2\omega^2}. \tag{10.23}$$

Since $\gamma \ll \omega$, the absorption cross section is sharply peaked near $\omega = \omega_0$. To a good approximation, $\omega^2 - \omega_0^2 = (\omega - \omega_0)(\omega + \omega_0)$ can be replaced by $2\omega_0(\omega - \omega_0)$, and $\omega$ can be replaced by $\omega_0$ everywhere else that it appears in equation (10.23) and also in equation (10.19) for the damping constant. Making this substitution in equation (10.23) and using equation (10.19), we obtain

$$\sigma(\omega) = \frac{\pi e^2}{m_ec}\frac{\gamma}{(\omega - \omega_0)^2 + \left(\dfrac{\gamma}{2}\right)^2}. \tag{10.24}$$

The integral of this expression over frequency is called the *total cross section* (but note that it has dimensions of area divided by time). It is useful as a normalization factor for the absorption cross section and is a measure of the strength of the absorption line. The integral is

$$\sigma_{\text{tot}} = \int_0^\infty \sigma(\omega) \mathrm{d}\nu = \frac{\pi e^2}{m_e c} \int_0^\infty \frac{\gamma \mathrm{d}(\omega/2\pi)}{(\omega - \omega_0)^2 + \left(\dfrac{\gamma}{2}\right)^2} = \frac{e^2}{m_e c} \int_{-\infty}^\infty \frac{\mathrm{d}x}{x^2 + 1} = \frac{\pi e^2}{m_e c}.$$

(10.25)

Hence

$$\frac{\sigma(\omega)}{\sigma_{\text{tot}}} = \frac{\gamma}{(\omega - \omega_0)^2 + \left(\dfrac{\gamma}{2}\right)^2}.$$

(10.26)

This is called a *Lorentz profile*.

The classical analysis predicts a unique scattering efficiency for all transitions. This is not surprising because it makes no reference to the actual atomic structure. A quantum mechanical treatment shows that the total cross sections for different transitions can differ by many orders of magnitude. A customary way of taking this into account is by introducing the *oscillator strength* of the transition, which is usually denoted by $f$. In a sense, $f$ is the effective number of classical oscillators involved in the transition.

In calculating the Rosseland mean opacity, line broadening must be accounted for. Line broadening arises from the Doppler shift associated with thermal motions of the absorbing particles and from interactions with other particles during the absorption of the photon.

## 10.8 The Rosseland mean opacity for solar composition material

In MS stars, the density varies with temperature roughly as $\rho \sim T^3$. Hence to obtain a rectangular grid of opacity data, it is more convenient to use, instead of density, a new variable $r = \rho/T_6^3$, where $T_6$ is the temperature in units of $10^6$ K. Contours of radiative opacity[1] are shown in figure 10.2 for near solar composition material with mass fractions $X = 0.7$, $Z = 0.02$.

The irregular cut out at the top right of figure 10.2 is where opacity data have not been calculated. The contours are labeled with the logarithm of the opacity in cgs units. We see that, in general, opacity increases with $r$ at constant $T$. If $r$ is kept fixed, we see that the opacity is low at low temperature, initially increases with temperature, reaches a maximum, and then declines as temperature increases further before reaching a plateau level. This behavior can be explained as follows. At low temperature, most of the elements are in their neutral atomic states. Only bound–bound transitions contribute to the opacity, and only weakly because only low lying

---

[1] The opacity data are from the OPAL opacity web site http://opalopacity.llnl.gov/. The relevant paper is [2].

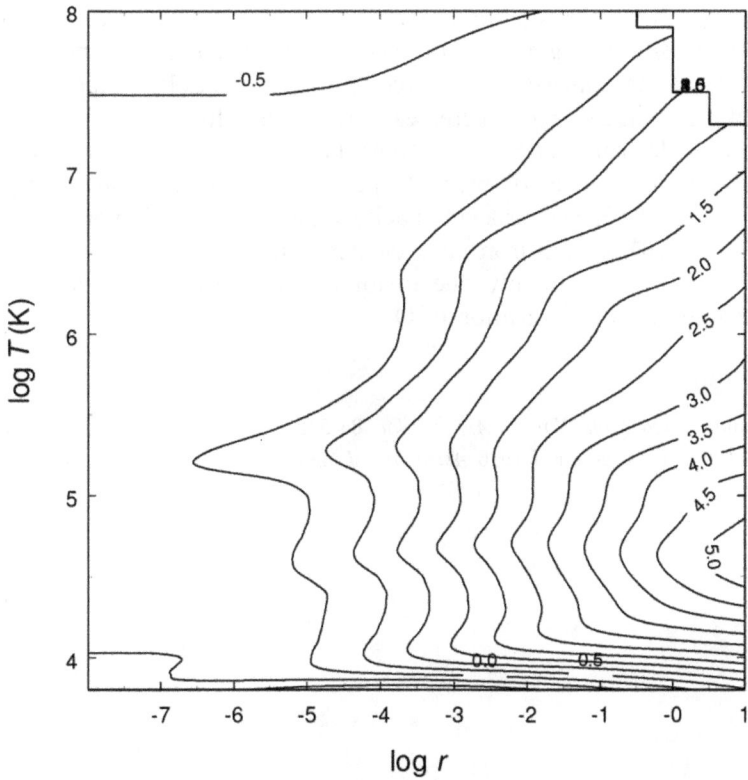

**Figure 10.2.** Contour plot of the logarithm of the Rosseland mean opacity for composition $X = 0.7$, $Z = 0.02$.

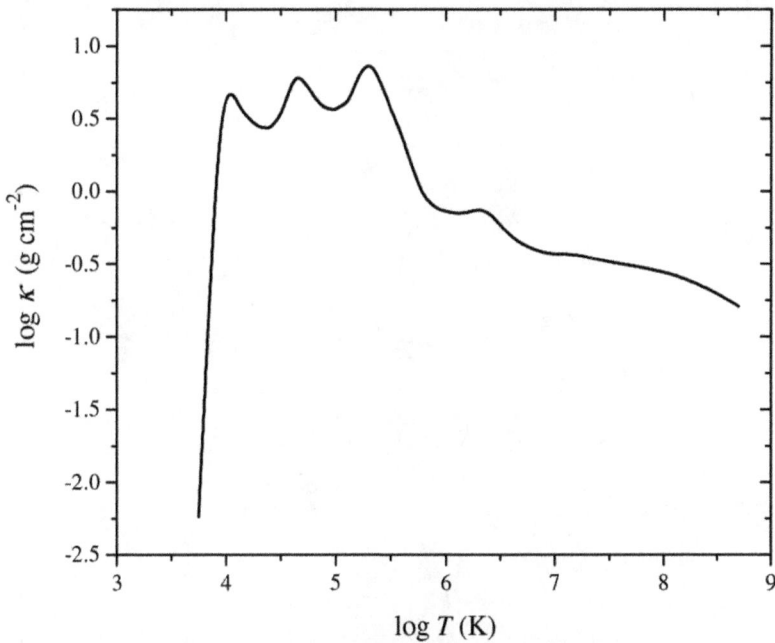

**Figure 10.3.** Opacity plotted against temperature at fixed $r$, for composition $X = 0.7$, $Z = 0.02$.

10-8

states are populated. At a slightly higher temperature, upper states are excited and bound–free transitions begin to occur. These give a steep increase in opacity. Once the major species are ionized, bound–free opacity becomes less important and free–free opacity dominates. This decreases with temperature and hence the opacity decreases until electron scattering becomes the dominant opacity source.

This behavior can be clearly seen when opacity is plotted against temperature at fixed $r$ as in figure 10.3. The peaks in opacity are due to ionization of H at $T = 10^4$ K, He at $T = 4 \times 10^4$ K, and iron peak elements at $T = 2.5 \times 10^5$ K. The peak at $T = 2 \times 10^6$ K is also mainly due to ionization of iron peak elements, with a small contribution from ionization of O.

## Bibliography

[1] Rosseland S 1924 *Mon. Not. R. Astron. Soc.* **84** 525
[2] Iglesias C A and Rogers F J 1996 *Astrophys. J.* **464** 943

# Chapter 11

## Nuclear reactions

## 11.1 Introduction

Atomic nuclei contain positively charged protons and uncharged neutrons. Since the Coulomb force between two protons is repulsive, there must be an attractive force acting in nuclei to keep them together. Since $^3$He and $^4$He are stable nuclei but the diproton ($^2$He) is unstable, we see that the range of the attractive force is of order the size of a nucleon ($\sim 10^{-13}$ cm). At this scale of separation between two protons, the Coulomb repulsion is quite large. Hence the strength of the attractive force at this scale must be high. This attractive force is aptly named the *strong nuclear force*.

The nuclear binding energy of an atom with $Z$ protons (and $Z$ electrons) and $N$ neutrons is

$$Q(Z, N) = \left[ Zm_p + Zm_e + Nm_n - m(A, Z) \right]c^2. \tag{11.1}$$

Here $m_p$ and $m_n$ are the masses of the proton and neutron, respectively and $m(A, Z)$ is the mass of the atom. The nucleon number is $A = Z + N$. (Note that it is conventional in nuclear physics to include the electron rest mass in $m(A, Z)$, so that it is the atomic mass rather than the nuclear mass. To obtain the nuclear mass, we must subtract off $Zm_e$.)

Figure 11.1 shows the nuclear binding energy per nucleon for naturally abundant isotopes as a function of proton number. We see that isotopes near Fe are more tightly bound (per nucleon) than both light nuclei and very heavy nuclei. Hence energy will be released if light nuclei are fused together.

The fusion of H into $^4$He releases $6.3 \times 10^{18}$ erg g$^{-1}$, whereas fusion of H into $^{56}$Fe would release $7.6 \times 10^{18}$ erg g$^{-1}$. (Complete annihilation of matter releases $9 \times 10^{20}$ erg g$^{-1}$.)

## 11.2 Occurrence of thermonuclear reactions

Because of the Coulomb repulsion between nuclei, nuclear fusion can only occur if the nuclei approach close enough that the strong force can come into play.

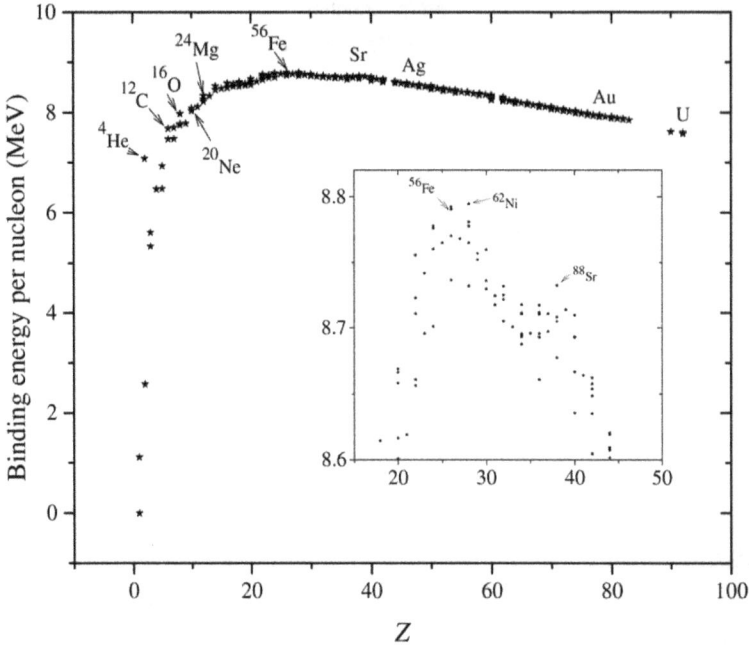

**Figure 11.1.** Binding energy per nucleon for naturally abundant isotopes. Data from [1].

The distance of closest approach of two nuclei of charge $Z_1$ and $Z_2$ and kinetic energy $kT$ is

$$d \sim \frac{Z_1 Z_2 e^2}{kT}. \tag{11.2}$$

For $d \sim 10^{-13}$ cm, we need $T \sim Z_1 Z_2 \, 10^{10}$ K. This is much higher than the central temperature of the Sun and shows that for particles to fuse they must have kinetic energy in the high energy tail of the Maxwellian distribution. However, the reaction rate would then be too low to provide the solar luminosity. The resolution to this dilemma lies in quantum mechanical tunneling through the Coulomb barrier.

## 11.3 Cross sections and nuclear reaction rates

Consider a situation in which a beam of particles $b$ is incident on a fixed target consisting of particles $Y$, as shown in figure 11.2.

Some of the particles in the beam fuse with particles in the target to form products $P$ and $d$,

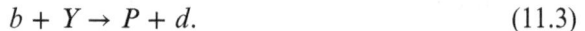

$$b + Y \rightarrow P + d. \tag{11.3}$$

The cross section for the reaction is

$$\sigma(v) = \frac{\text{number of reactions per particle } Y \text{ per unit time}}{\text{number of incident particles } b \text{ per unit area per unit time}}. \tag{11.4}$$

This is a function of the relative velocity, $v$, of particles $b$ and $Y$.

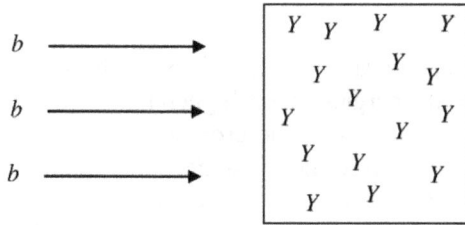

**Figure 11.2.** Schematic of a beam of particles incident on a fixed target.

The number of reactions per unit time per unit volume is

$$r_{bY} = \sigma(v)vN_bN_Y, \tag{11.5}$$

where $N_b$ and $N_Y$ are particle number densities. This follows from the definition (11.4).

In stars, there will be a distribution of relative velocities of the interacting particles. Let this distribution be $\phi(v)$, with normalization $\int_0^\infty \phi(v)dv = 1$. The reaction rate in stars is then obtained by averaging (11.5) over the velocity distribution

$$r_{bY} = N_bN_Y\int_0^\infty \sigma(v)v\phi(v)dv = N_bN_Y\langle\sigma v\rangle. \tag{11.6}$$

In this derivation, we have assumed that $b$ and $Y$ are different kinds of nuclei. For identical particles, the total number of pairs of particles is not $N_bN_Y$ but $N_b^2/2$. Hence to include this case equation (11.6) is modified to

$$r_{bY} = \frac{N_bN_Y\langle\sigma v\rangle}{1 + \delta_{bY}}, \tag{11.7}$$

where

$$\delta_{bY} = \begin{cases} 1 & \text{if } b = Y \\ 0 & \text{if } b \neq Y. \end{cases} \tag{11.8}$$

In thermal equilibrium, the velocity distributions for all the different species of particles will be Maxwellian with the same temperature, $T$. The relative velocity distribution for particles of species 1 and 2 will also be Maxwellian with an effective mass equal to the reduced mass of the two species. The reaction rate per pair of particles is then

$$\lambda = \langle\sigma v\rangle = 4\pi\left(\frac{\mu}{2\pi kT}\right)^{3/2}\int_0^\infty v^3\sigma(v)\exp\left(-\frac{\mu v^2}{2kT}\right)dv, \tag{11.9}$$

where the reduced mass is

$$\mu = \frac{m_1m_2}{m_1 + m_2}. \tag{11.10}$$

## 11.4 The cross section

In principle, the cross section can be measured by experiment. However, in practice, this can be done only at energies much higher than are relevant for most fusion reactions in stars. Extrapolation of the cross section to lower energies is needed. Blindly extrapolating over a large energy range is very likely to lead to large errors in the cross section and reaction rates. To reduce the error, guidance from theoretical understanding of what determines the cross section is used. An important part of this is the probability of quantum mechanical tunneling through the Coulomb barrier. This is shown schematically in figure 11.3.

Shown here is the interaction potential that consists of the repulsive Coulomb potential and the attractive potential associated with the strong nuclear force of range $r_s$. A typical center-of-mass energy relevant to thermonuclear reactions under stellar conditions, $E_n$, is also shown together with its classical turning radius, $R$. For fusion to occur the particles must tunnel from $R$ through the Coulomb barrier of height, $V_B$, to radius $r_s$. A quantum mechanical calculation [2] gives that, for $E_n \ll V_B$, the tunneling probability is

$$W = \exp(-2G), \tag{11.11}$$

where

$$G = (2\mu)^{1/2} \frac{\pi^2 Z_1 Z_2 e^2}{h E^{1/2}}. \tag{11.12}$$

Here the energy $E$ is related to the relative velocity by

$$E = \frac{1}{2}\mu v^2. \tag{11.13}$$

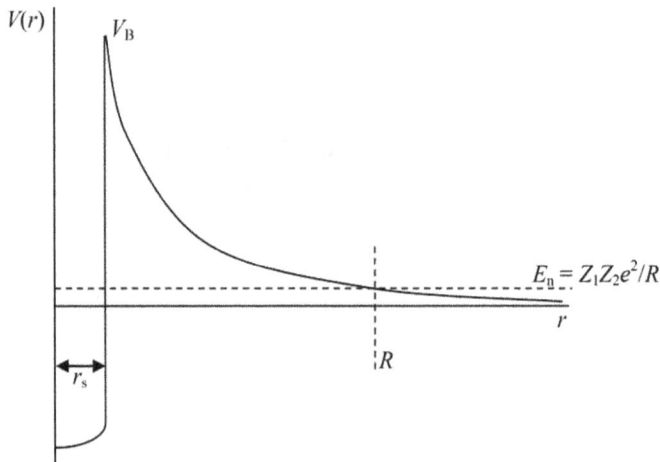

**Figure 11.3.** The Coulomb barrier.

Due to the Heisenberg uncertainty principle, the cross section has a geometrical factor that is inversely proportional to the energy. The uncertainty, $\Delta r$, in the position of a particle of momentum $p$ is

$$\Delta r \sim \frac{\hbar}{p} = \frac{\hbar}{(2\mu E)^{1/2}}. \tag{11.14}$$

This is a measure of the effective size of the particle. Hence

$$\sigma \propto (\Delta r)^2 \sim \frac{\hbar^2}{2\mu E} \propto \frac{1}{E}. \tag{11.15}$$

Once the tunneling probability and geometrical factor are taken out of the cross section, we are left with the nuclear cross section factor, $S(E)$, which represents the intrinsically nuclear contributions to the cross section. Specifically,

$$\sigma(E) = \frac{S(E)}{E} e^{-2G}. \tag{11.16}$$

Unless there is a resonance (i.e. the target nucleus has an internal state of energy comparable to the energy of the incident particle), $S(E)$ is a slowly varying function of $E$ and hence extrapolation of $S(E)$ to low energies is not expected to lead to significant errors in reaction rates.

## 11.5 Evaluation of the reaction rate

In terms of $E$, the integral expression for the reaction rate per pair of particles is

$$\lambda = \langle \sigma v \rangle = \left(\frac{8}{\pi\mu}\right)^{1/2} \left(\frac{1}{kT}\right)^{3/2} \int_0^\infty S(E) \exp\left(-\frac{E}{kT} - \frac{b}{E^{1/2}}\right) dE, \tag{11.17}$$

where

$$b = \frac{(8\mu)^{1/2}\pi^2 Z_1 Z_2 e^2}{h} = 31.28 Z_1 Z_2 A^{1/2} \ (\text{keV})^{1/2}. \tag{11.18}$$

Here

$$A = \frac{A_1 A_2}{A_1 + A_2} = \frac{\mu}{m_u}. \tag{11.19}$$

Since $S(E)$ is a slowly varying function of $E$, we see that the energies which give the dominant contribution to the integral are those near where the argument of the exponential is stationary. This occurs at energy

$$E_0 = \left(\frac{bkT}{2}\right)^{2/3} = 1.220\left(Z_1^2 Z_2^2 A T_6^2\right)^{1/3} \ \text{keV}, \tag{11.20}$$

where $T_6$ is the temperature in units of $10^6$ K.

For fusion of hydrogen at the solar center, $E_0 = 6$ keV. This is higher than the typical particle energy of 1 keV but much less than the energy needed for fusion in the absence of quantum tunneling, which is of order 1 MeV.

The integral in equation (11.17) is usually performed by some variation of the method of steepest descent. The argument of the exponential term is first expanded in a Taylor series about its stationary point,

$$-\frac{E}{kT} - \frac{b}{E^{1/2}} = -\frac{E_0}{kT} - \frac{b}{E_0^{1/2}} - \frac{3}{8}\frac{b}{E_0^{5/2}}(E - E_0)^2 + \cdots = -\frac{3E_0}{kT} - \frac{3E_0}{kT}\frac{(E - E_0)^2}{4E_0^2} + \cdots$$

(11.21)

Truncating the series at the second term and replacing $S(E)$ by $S_0 = S(E_0)$, we have

$$\lambda \approx \left(\frac{8}{\pi\mu}\right)^{1/2}\left(\frac{1}{kT}\right)^{3/2} S(E_0) \int_{-\infty}^{\infty} \exp\left(-\frac{3E_0}{kT} - \frac{3E_0}{kT}\frac{(E - E_0)^2}{4E_0^2}\right) dE$$

$$= \frac{2}{3}\left(\frac{8}{3\pi\mu E_0}\right)^{1/2} S_0 \tau^2 \exp(-\tau) \int_{-\infty}^{\infty} \exp\left(-x^2\right) dx$$

$$= \frac{7.20 \times 10^{-19}}{AZ_1Z_2} \frac{S_0}{1\text{ keV barn}} \tau^2 \exp(-\tau) \text{ cm}^3 \text{ s}^{-1},$$

(11.22)

where

$$\tau = \frac{3E_0}{kT} = 42.48\left(\frac{Z_1^2 Z_2^2 A}{T_6}\right)^{1/3}.$$

(11.23)

Note that the barn is a unit of area equal to $10^{-24}$ cm$^2$. It is a convenient unit to measure nuclear cross sections.

The reaction rate is

$$r_{12} = \frac{2.62 \times 10^{17}}{(1 + \delta_{12})AZ_1Z_2} \frac{S_0}{1\text{ keV barn}} \frac{X_1 X_2}{A_1 A_2} \rho^2 \tau^2 \exp(-\tau) \text{ cm}^{-3} \text{ s}^{-1},$$

(11.24)

where density is in units of g cm$^{-3}$.

Note that this formula is approximate and more accurate formulae have been developed to take into account the variation of the nuclear cross section factor with energy, including the presence of resonances (see e.g. [3]).

From equation (11.24), we have

$$\frac{\partial \ln r_{12}}{\partial \ln T} = \frac{\partial}{\partial \ln T}(2 \ln \tau - \tau) = -\frac{2}{3} + 14.16\left(\frac{Z_1^2 Z_2^2 A}{T_6}\right)^{1/3}.$$

(11.25)

We see that at low temperatures, the reaction rate increases steeply with temperature, but at a high enough temperature the reaction rate actually decreases with temperature.

## 11.6 Major nuclear burning stages in stars: H burning

In stars, the first nuclear reaction that occurs is deuterium burning in which the deuterium produced originally in the big bang is destroyed by the reaction

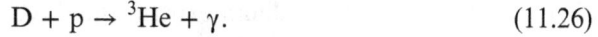

$$D + p \rightarrow {}^3He + \gamma. \tag{11.26}$$

Because the primordial D abundance is $X_D \sim 2 \times 10^{-5}$, in most stars the deuterium phase is relatively short lived. However deuterium burning is the only thermal equilibrium nuclear energy burning phase for brown dwarf stars which have masses in the range $\sim 0.013$ to $\sim 0.072$ $M_\odot$. In stars more massive than $0.072$ $M_\odot$, deuterium burning is immediately followed by the major core hydrogen burning phase. The specific reactions involved in fusing four protons to form a helium nucleus depend on the central temperature, which is determined by the mass of the star. In stars of mass less than about $1.2$ $M_\odot$, the main reactions are the pp-chains. These are a set of three linked reaction chains, each ending in the formation of helium nuclei.

The reactions of the ppI chain are

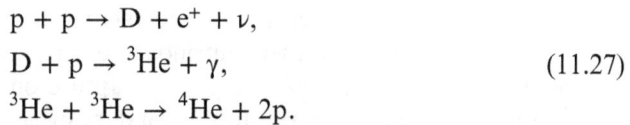

$$p + p \rightarrow D + e^+ + \nu,$$
$$D + p \rightarrow {}^3He + \gamma, \tag{11.27}$$
$${}^3He + {}^3He \rightarrow {}^4He + 2p.$$

Because it involves a weak interaction, the first of these reactions is by far the slowest and determines the rate of the pp chains. The second reaction is much faster than the first reaction, and so the deuterium nuclei are barely formed before they are destroyed. Hence the deuterium abundance does not build up to its primordial level. The neutrino produced by the first reaction escapes from the star and carries away on average 0.263 MeV.

The reactions of the ppII chain are

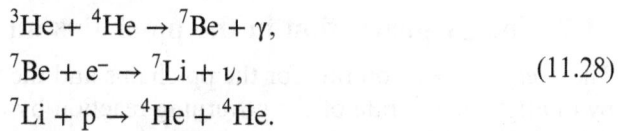

$${}^3He + {}^4He \rightarrow {}^7Be + \gamma,$$
$${}^7Be + e^- \rightarrow {}^7Li + \nu, \tag{11.28}$$
$${}^7Li + p \rightarrow {}^4He + {}^4He.$$

The neutrino produced in the second reaction carries away 0.80 MeV on average.

The reactions of the ppIII chain are

$${}^7Be + p \rightarrow {}^8B + \gamma,$$
$${}^8B \rightarrow {}^8Be + e^+ + \nu, \tag{11.29}$$
$${}^8Be \rightarrow {}^4He + {}^4He.$$

The neutrino produced in the second reaction of the ppIII chain carries away 7.2 MeV on average. These are the neutrinos detected by the Homestake neutrino experiment [4], which consisted of a 100 000 gallon tank of perchlorethylene ($C_2Cl_4$). Only the ${}^8B$ neutrinos are sufficiently energetic to convert ${}^{37}Cl$ nuclei into ${}^{37}Ar$ nuclei. Because the neutrinos are very weakly interacting, only a few ${}^{37}Ar$ nuclei were produced during each month of operation. Surprisingly, the neutrino production rate was about a factor three less than predicted by solar models. This is now attributed to neutrino flavor oscillations in which some of the electron neutrinos change into muon neutrinos and tau neutrinos on the way to Earth. More recent experiments such as the gallium experiments SAGE and GALLEX have detected the much more numerous but less energetic neutrinos from the p + p reaction.

Figure 11.4 shows how the central values of the mass fractions of some of the light nuclei involved in the pp chains evolve with time for a star of mass 1 $M_\odot$ and heavy-element abundance $Z = 0.017$.

From the right-hand panel, we see that the deuterium burning phase only lasts about 200 000 years.

In stars more massive than the Sun, core hydrogen burning proceeds through the CNO-cycles. The main cycle is

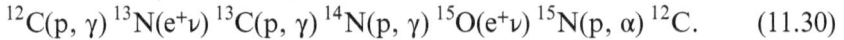

$$^{12}C(p, \gamma)\ ^{13}N(e^+\nu)\ ^{13}C(p, \gamma)\ ^{14}N(p, \gamma)\ ^{15}O(e^+\nu)\ ^{15}N(p, \alpha)\ ^{12}C. \qquad (11.30)$$

Note that the cycle begins and ends with a $^{12}C$ nucleus. Once the cycle reaches equilibrium, the CNO nuclei act as catalysts to fuse four protons into one alpha particle plus two positrons, two neutrinos and three photons.

About four times in 10 000 the proton capture on $^{15}N$ results in the formation of a $^{16}O$ nucleus. This gives rise to a secondary cycle

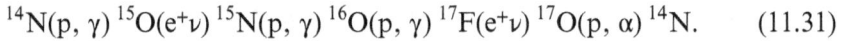

$$^{14}N(p, \gamma)\ ^{15}O(e^+\nu)\ ^{15}N(p, \gamma)\ ^{16}O(p, \gamma)\ ^{17}F(e^+\nu)\ ^{17}O(p, \alpha)\ ^{14}N. \qquad (11.31)$$

The slowest reaction in the CNO-cycles is the proton capture on $^{14}N$. Hence in the burning region of the star, the original C and O nuclei are converted mainly to $^{14}N$. Since C and O are more abundant than N in the Sun's surface layers and in the solar system in general, in situations where nuclear processed material reaches the stellar surface, a high N to C ratio is taken as an indication that CNO cycling has occurred.

## 11.7 Energy generation in the pp-chains and the CNO-cycles

The energy generation rate for the pp-chains and the CNO-cycles can be estimated by multiplying the rate of the governing reaction by the energy released, excluding that carried off by neutrinos.

The slowest reaction for the pp-chains is the p + p reaction for which from equation (11.24), with $S_0 = 4.00 \times 10^{-22}$ keV barn from experiment,

$$r_{pp} = 1.05 \times 10^8 X_H^2 \rho^2 \tau^2 \exp(-\tau)\ \text{cm}^{-3}\,\text{s}^{-1}, \qquad (11.32)$$

where

$$\tau = 33.72 T_6^{-1/3}. \qquad (11.33)$$

The neutrino energy loss depends on which particular pp chains are involved in producing the $^4$He nuclei. For the ppI chain the energy produced per unit mass per unit time is

$$\varepsilon_{ppI} = 2.3710^9 X_H^2 \rho \exp\left(-33.72 T_6^{-1/3}\right)\big/ T_6^{2/3}\ \text{erg g}^{-1}\,\text{s}^{-1}. \qquad (11.34)$$

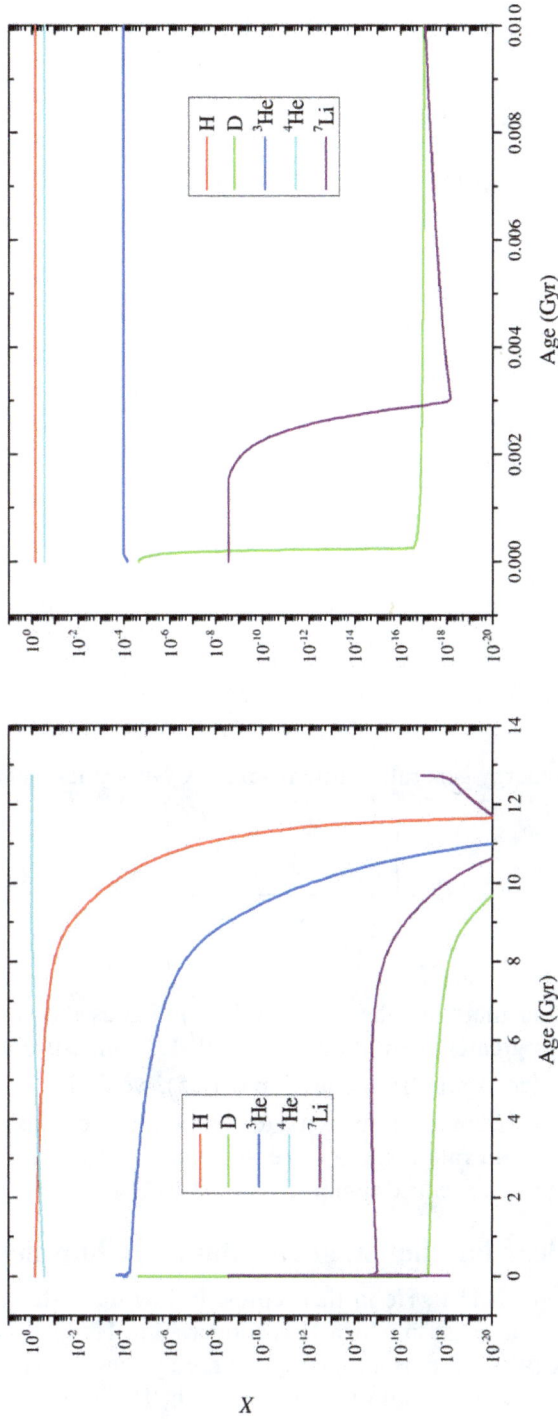

**Figure 11.4.** Evolution of the central mass fractions of light elements for a $Z = 0.017$, $1\ M_\odot$ model.

The ppII and ppIII chains can be included by multiplying this by correction factor

$$\varepsilon_{pp} = \left(F_{ppI} + 0.980 F_{ppII} + 0.736 F_{ppIII}\right)\varepsilon_{ppI}, \qquad (11.35)$$

where $F_{ppI}$, $F_{ppII}$, and $F_{ppIII}$ are the fractions of alpha particles made by the ppI, ppII, and ppIII chains, respectively. From equation (11.25), we find for the pp chains at the solar center, $\partial \ln r_{pp}/\partial \ln T \approx 4$.

The slowest reaction in the CNO-cycles is the $^{14}$N proton capture, which has $S_0 = 3.2$ keV barn.

Its rate is

$$r_{14p} = 9.15 \times 10^{27} X_H X_{14} \rho^2 \tau^2 \exp(-\tau) \, \text{cm}^{-3}\,\text{s}^{-1}, \qquad (11.36)$$

where now

$$\tau = 152.3 T_6^{-1/3}. \qquad (11.37)$$

In equilibrium, almost all of the original CNO nuclei will have been converted to $^{14}$N, and hence we can write the energy generation rate as

$$\varepsilon_{CNO} = 8 \times 10^{30} X_H X_{CNO} \rho \exp\left(-152.31 T_6^{-1/3}\right)/T_6^{2/3} \, \text{erg g}^{-1}\,\text{s}^{-1}. \qquad (11.38)$$

The ratio of the rate of energy generation by the CNO-cycles to that from the ppI chain is

$$\frac{\varepsilon_{CNO}}{\varepsilon_{ppI}} = 3.4 \times 10^{21} \frac{X_{CNO}}{X_H} \exp\left(-118.6 T_6^{-1/3}\right), \qquad (11.39)$$

which gives that the energy generation rate from the CNO-cycles will be the greater if

$$T_6 > \left[\frac{118.6}{49.6 + \ln\left(\dfrac{X_{CNO}}{X_H}\right)}\right]^3. \qquad (11.40)$$

For solar composition material, the CNO-cycle dominates the energy generation rate for temperatures greater than about $1.8 \times 10^7$ K, which is about 20% higher than at the Sun's center. Again from equation (11.25), we find $\frac{\partial \ln r_{CNO}}{\partial \ln T} \approx 19$. As we will see later, this much greater sensitivity to temperature of the CNO-cycle rate compared to the pp-chain rate is a major reason that stars more massive than the Sun are convective in their central regions.

## 11.8 Major nuclear burning stages in stars: He burning

On the MS, stars convert H to He in their cores. In low mass stars, core hydrogen burning is followed by a phase in which H is converted to He in a shell surrounding the He core. During the shell burning phase, the core grows in mass, shrinks in size and heats up. When the temperature is high enough, He burning reactions begin. Due to the larger Coulomb repulsion between the He nuclei, the required temperature is higher than for the pp reaction. In massive stars, core hydrogen burning is almost immediately followed by core helium burning.

The first set of reactions during helium burning is called the triple-alpha (3α) reaction, in which three alpha particles fuse to make a $^{12}$C nucleus. A major hurdle in understanding how the 3α reaction proceeds is that the $^{8}$Be nucleus formed from fusing two α particles is unstable (to break up into two α particles). However $^{8}$Be is unstable by a relatively small energy of 92 keV. The $3 \times 10^{-16}$ s lifetime of $^{8}$Be is longer than the time for two α particles to scatter past each other. Hence $^{8}$Be builds up to a small equilibrium abundance ($\sim 10^{-9}$), which is sufficient for a third alpha to be captured. The reactions are

$$\begin{aligned} ^{4}\text{He} + {}^{4}\text{He} &\rightleftarrows {}^{8}\text{Be}, \\ ^{8}\text{Be} + {}^{4}\text{He} &\rightarrow {}^{12}\text{C} + \gamma. \end{aligned} \qquad (11.41)$$

Once a sufficient abundance of $^{12}$C has been built up, the reaction

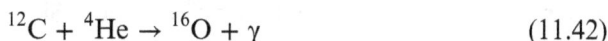

$$^{12}\text{C} + {}^{4}\text{He} \rightarrow {}^{16}\text{O} + \gamma \qquad (11.42)$$

becomes important and proceeds together with the 3α reaction. Similarly the reactions

$$^{16}\text{O} + {}^{4}\text{He} \rightarrow {}^{20}\text{Ne} + \gamma \qquad (11.43)$$

and

$$^{20}\text{Ne} + {}^{4}\text{He} \rightarrow {}^{24}\text{Mg} + \gamma \qquad (11.44)$$

can occur.

Calculations indicate that at the end of core helium burning, the composition at the center of a 1 $M_\odot$ Pop I star is mainly 31% $^{12}$C, 67% $^{16}$O, and 2% $^{24}$Mg by mass. After core helium burning has ended, a low mass star will experience a shell helium burning phase during which the carbon–oxygen (CO) core grows in mass. Observations of young open clusters such as the Pleiades and NGC 2516 indicate that single stars of initial mass less than about 7 $M_\odot$ end their lives as WDs. Hence these stars must lose sufficient mass that the core cannot exceed the Chandrasekhar mass limit. Stellar evolution models show that if the initial mass of the star is less than about 8 $M_\odot$, the core does not become hot enough for the next fuel, $^{12}$C, to ignite [5, 6]. The star ends its life as a CO WD.

## 11.9 Advanced nuclear burning phases

After helium burning in stars initially more massive than about 8 $M_\odot$, gravitational contraction heats the core enough that carbon burning starts. The main reaction is

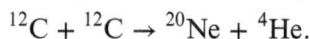

$$^{12}\text{C} + {}^{12}\text{C} \rightarrow {}^{20}\text{Ne} + {}^{4}\text{He}.$$

During carbon burning the released α particles react with $^{12}$C, $^{16}$O, and to a lesser extent with other nuclei. At the end of carbon burning the core is mostly $^{16}$O, $^{20}$Ne, and $^{24}$Mg.

If the star has an initial mass greater than about 12 $M_\odot$, the core experiences a sequence of burning phases that convert it into iron peak elements. The first of these burning phases is called neon burning. This is initiated by a reverse (photo-disintegration) reaction

$$^{20}\text{Ne} + \gamma \rightarrow {}^{16}\text{O} + {}^{4}\text{He}.$$

The released $\alpha$ particles react with $^{20}$Ne and $^{24}$Mg,

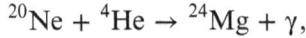

$$^{20}\text{Ne} + {}^{4}\text{He} \rightarrow {}^{24}\text{Mg} + \gamma,$$
$$^{24}\text{Mg} + {}^{4}\text{He} \rightarrow {}^{28}\text{Si} + \gamma.$$

At the end of neon burning the core is mostly $^{16}$O, $^{24}$Mg, and $^{28}$Si.

This is followed by oxygen burning. The dominant reaction is

$$^{16}\text{O} + {}^{16}\text{O} \rightarrow {}^{28}\text{Si} + {}^{4}\text{He}.$$

The released $\alpha$ particles react with $^{24}$Mg and $^{28}$Si,

$$^{24}\text{Mg} + {}^{4}\text{He} \rightarrow {}^{28}\text{Si} + \gamma,$$
$$^{28}\text{Si} + {}^{4}\text{He} \rightarrow {}^{32}\text{S} + \gamma.$$

At the end of oxygen burning the core is mainly $^{28}$Si and $^{32}$S. Because of the large Coulomb barrier, the reactions $^{28}$Si $+ {}^{28}$Si, $^{28}$Si $+ {}^{32}$S do not occur. Instead, the evolution proceeds through photo-disintegration of less tightly bound nuclei and capture of liberated light particles (p, n, $\alpha$) to steadily build up heavier and heavier tightly bound nuclei. The end result of *silicon burning* is that the core consists of iron peak elements, mainly $^{56}$Fe. The reason that the core is mainly $^{56}$Fe instead of the more tightly bound $^{62}$Ni is that there is a sequence of alpha capture reactions that naturally leads to $^{56}$Ni, which is unstable and decays to $^{56}$Fe.

## Bibliography

[1] Audi G and Wapstra A H 1993 *Nucl. Phys.* A **565** 1
[2] Gamow G 1928 *Z. Phys.* **51** 204
[3] Angulo C *et al* 1999 *Nucl. Phys.* A **656** 3
[4] Cleveland B T *et al* 1998 *Astrophys. J.* **496** 505
[5] García-Berro E *et al* 1997 *Astrophys. J.* **485** 765
[6] Doherty C L, Gil-Pons P, Siess L, Lattanzio J C and Lau H H B 2015 *Mon. Not. R. Astron. Soc.* **446** 2599

# Chapter 12

## Neutrino energy loss processes

In addition to the neutrinos produced in nuclear reactions, there are a number of purely leptonic processes in which a pair of neutrinos can be emitted when an electron changes its momentum. Because neutrinos interact very weakly with matter, they usually pass through the stellar material into space carrying away energy from the star. The processes described in the following sections can be important in stellar interiors [1].

### 12.1 Pair annihilation neutrino process $(e^+ + e^- \rightarrow \nu + \bar{\nu})$

In very hot environments $(T > 10^9$ K), electron–positron pairs can be created by photon processes. The electron–positron pairs are soon annihilated, usually giving two photons but once in about $10^{19}$ cases a neutrino pair $(\nu\bar{\nu})$ is produced. If the plasma is not too dense, the neutrinos escape from the star without interaction. The energy loss rate is a complicated function of temperature and density but there are simple limiting cases for non-degenerate electrons:

$$\rho\varepsilon_\nu = \begin{cases} 4.9 \times 10^{18} T_9^3 \exp\left(-\dfrac{11.86}{T_9}\right), & T_9 < 1, \\ 4.6 \times 10^{15} T_9^9, & T_9 > 3. \end{cases} \tag{12.1}$$

Here the energy loss rate has units of erg g$^{-1}$ s$^{-1}$ and $T_9 = T/10^9$ K.

Electron degeneracy reduces the neutrino loss rate by reducing the amount of phase space available for electron–positron pair production.

### 12.2 Plasma neutrino process $(\gamma_{\text{plasmon}} \rightarrow \nu + \bar{\nu})$

A single photon cannot decay into a neutrino pair unless the neutrinos move in opposite directions. (This is because the photon has spin 1 and the neutrinos are spin ½ particles of opposite helicity.) If the neutrinos move in opposite directions, the decay cannot take place in a vacuum because energy and momentum cannot be simultaneously conserved. The decay is possible when the photon moves through plasma. The *dispersion relation* that relates angular frequency, $\omega$ and wave number, $k$, for electromagnetic waves propagating in plasma is

$$\omega^2 = k^2c^2 + \omega_p{}^2, \tag{12.2}$$

where $\omega_p$ is the plasma frequency. If the electrons in the plasma are non-degenerate, then

$$\omega_p{}^2 = \frac{4\pi e^2 n_e}{m_e}. \tag{12.3}$$

This expression is modified by electron degeneracy to

$$\omega_p{}^2 = \frac{4\pi e^2 n_e}{m_e}\left[1 + \left(\frac{\hbar}{m_e c}\right)^2 (3\pi^2 n_e)^{2/3}\right]^{-1/2}. \tag{12.4}$$

At low densities, $\hbar\omega_p$ is small compared to the typical photon energy $\simeq kT$, and photon processes are largely unaffected by the presence of plasma. The situation is different in high density electron-degenerate stellar interiors where $\hbar\omega_p$ can be comparable to $kT$.

By comparing the energy-momentum relation for a massive particle

$$E^2 = p^2c^2 + m^2c^4, \tag{12.5}$$

to the dispersion relation (12.2), we see that an electromagnetic wave in plasma is kinematically equivalent to a relativistic particle. The quantized wave is called a *plasmon*, and has an equivalent rest mass

$$m_{\text{plasmon}} = \frac{\hbar\omega_p}{c^2}. \tag{12.6}$$

It is the plasmon rest energy $\hbar\omega_p$ that allows energy and momentum to be conserved in the decay of a photon into a neutrino pair.

With the definitions

$$x = \frac{\hbar\omega_p}{kT}, \tag{12.7}$$

and

$$y = \frac{kT}{m_e c^2}, \tag{12.8}$$

the plasma neutrino loss rate has the limiting forms

$$\rho\varepsilon_\nu = \begin{cases} 7.4 \times 10^{21}\, y^3 x^6, & x \ll 1, \\ 3.85 \times 10^{21}\, y^9 x^{15/2} e^{-x}, & x \gg 1. \end{cases} \tag{12.9}$$

## 12.3 Photo-neutrino process ($\gamma + e \rightarrow e + \nu + \bar{\nu}$)

This process is the analog of Compton scattering. The outgoing photon is replaced by a neutrino pair. This process competes with the pair annihilation process only at temperatures that are so low that electron–positron pairs are not created, and it competes with the plasma neutrino process only at densities so low that the plasmon rest mass is trivially small. Limiting forms for the energy loss rate are

$$\rho \varepsilon_\nu = \begin{cases} 0.98 \times 10^8 \dfrac{\rho}{\mu_e} T_9^8, & \text{non-relativistic non-degenerate} \\[2ex] 4.8 \times 10^{11} \left( \dfrac{\rho}{\mu_e} \right)^{1/3} T_9^9, & \text{non-relativistic degenerate.} \end{cases} \tag{12.10}$$

## 12.4 Bremsstrahlung neutrino process

This process is the analog of free–free emission. The outgoing photon is replaced by a neutrino pair. At high densities the energy loss rate is

$$\varepsilon_\nu = 7.6 \times 10^5 \frac{Z^2}{A} T_9^6, \tag{12.11}$$

where $Z$ and $A$ are the charge and mass number of the nucleus.

Figure 12.1 is a contour plot of the neutrino loss rate per unit mass in units of erg g s$^{-1}$. Figure 12.2 shows the dominant neutrino loss process in the log$\rho$–log$T$ plane.

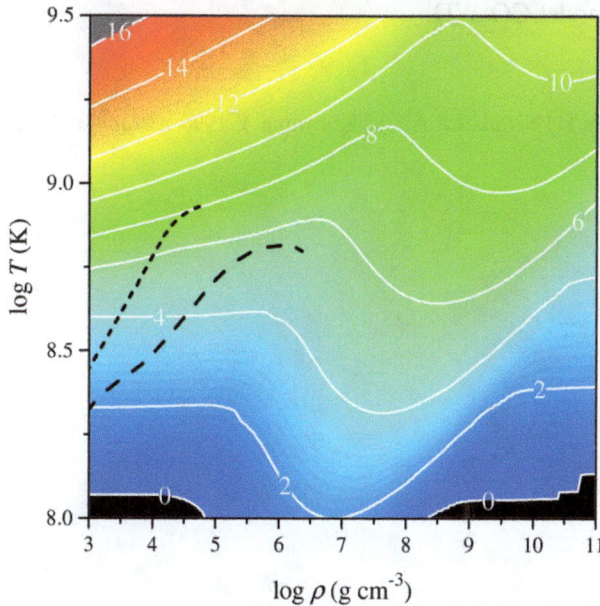

**Figure 12.1.** Contours of the neutrino loss rate per unit mass in units of erg g s$^{-1}$. The broken lines show the run of temperature with density in the central regions of 10 and 50 $M_\odot$ Pop I stars at the time of carbon ignition.

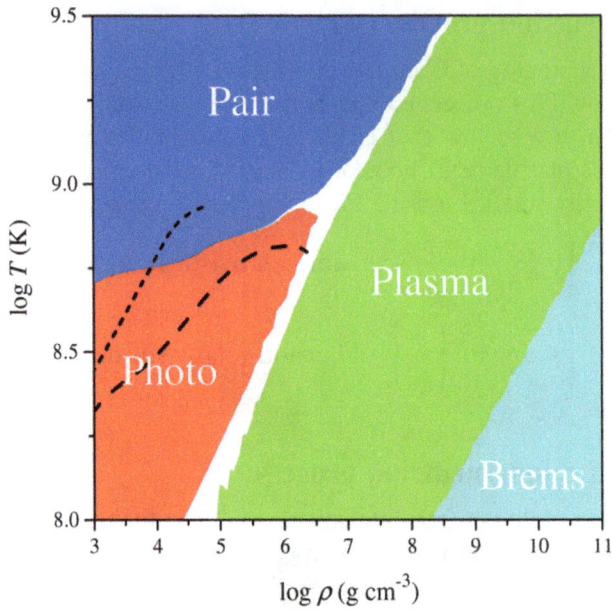

**Figure 12.2.** Regions of the $\log\rho$–$\log T$ plane where a neutrino loss process is dominant. The broken lines show the run of temperature with density in the central regions of 10 and 50 $M_\odot$ Pop I stars at the time of carbon ignition.

The broken lines show the run of temperature with density in the central regions of 10 and 50 $M_\odot$ Pop I stars at the time of carbon ignition. The composition is that at the center of a model CO WD.

## Bibliography

[1] Itoh N, Hayashi H, Nishikawa A and Kohyama Y 1996 *Astrophys. J. Suppl.* **102** 411

# Chapter 13

## Homology relations

### 13.1 Introduction

Stars of similar mass and composition are expected to have similar physical conditions in their interiors and hence are expected to have a similar structure. Homology relations provide a very approximate but still useful way of investigating how the stellar structure depends on the stellar mass and composition.

### 13.2 Homology of zero age main sequence stars

Consider a star on the ZAMS. It will be in thermal equilibrium and will have essentially uniform composition. Depending on the star's mass, it will have convection zones in the center or in the surface layers. For the time being, we will ignore the complications that arise from convective energy transport and will assume that the star is radiative throughout. Using the mass as the independent variable, the equations of stellar structure are

$$\frac{dr}{dm} = \frac{1}{4\pi r^2 \rho}, \tag{13.1}$$

$$\frac{dp}{dm} = -\frac{Gm}{4\pi r^4}, \tag{13.2}$$

$$\frac{dL}{dm} = \varepsilon, \tag{13.3}$$

and

$$\frac{dT}{dm} = -\frac{3\kappa L}{64\pi^2 acr^4 T^3}. \tag{13.4}$$

In each equation, the right-hand side is a product of a number of quantities. This suggests that we might be able to find *homology relations* that describe how the dependent variables scale with the total mass of the star, $M$. However, this will be possible only if the relations for $\varepsilon$, $\kappa$, $\rho$ in terms of the dependent variables are also multiplicative in nature. Fortunately, in many situations, this is approximately the case.

Often we are most interested in quantities at the stellar center or at the stellar surface. To derive the homology relations, we need as an independent variable a quantity that gives the location of the center and surface in a way that is independent of $M$. Such a variable is the scaled mass,

$$q = \frac{m}{M}. \tag{13.5}$$

The center of the star is at $q = 0$ and the surface is at $q = 1$.

We now assume that the dependent variables are of form

$$\begin{aligned}
r &= M^{a_r}\tilde{r}(q), \\
p &= M^{a_p}\tilde{p}(q), \\
L &= M^{a_L}\tilde{L}(q), \\
T &= M^{a_T}\tilde{T}(q),
\end{aligned} \tag{13.6}$$

where the exponents $a_r$, $a_p$, $a_L$, and $a_T$ are all constants, and $\tilde{r}(q)$, $\tilde{p}(q)$, $\tilde{L}(q)$, and $\tilde{T}(q)$ are functions only of $q$.

From equation (9.2), we obtain

$$\frac{M^{a_r}}{M}\frac{d\tilde{r}}{dq} = \frac{1}{4\pi M^{2a_r}\tilde{r}^2\rho}, \tag{13.7}$$

so that

$$\rho = \frac{1}{M^{3a_r-1}}\frac{1}{4\pi\tilde{r}^2}\left(\frac{d\tilde{r}}{dq}\right)^{-1}. \tag{13.8}$$

Hence the scaling of the density with $M$ is

$$\rho \propto \frac{1}{M^{3a_r-1}}. \tag{13.9}$$

From equation (13.2), we find

$$\frac{M^{a_p}}{M}\frac{d\tilde{p}}{dq} = -\frac{M}{M^{4a_r}}\frac{Gq}{4\pi\tilde{r}^4}, \tag{13.10}$$

so that

$$M^{4a_r+a_p-2} = -\frac{Gq}{4\pi\tilde{r}^4}\left(\frac{d\tilde{p}}{dq}\right)^{-1}. \tag{13.11}$$

Since the right-hand side is independent of $M$, we find that

$$4a_r + a_p - 2 = 0. \tag{13.12}$$

Without going into the details, from equations (13.3) and (13.4), we find that the nuclear energy generation rate and opacity scale with $M$ as

$$\varepsilon \propto M^{a_L-1}, \tag{13.13}$$

and

$$\kappa \propto M^{4a_r+4a_T-a_L-1}. \tag{13.14}$$

The relations in equations (13.9), (13.12), (13.13), and (13.14) are quite general because we have not yet made use of constitutive relations, i.e. the equation of state, opacity law, or energy generation rate. We have one relation between the four homology exponents. We can find three more relations by specifying the constitutive relations. There are a number of simple yet physically relevant possibilities for these relations depending on the physical conditions in the stellar interior. For example for a low mass star, it is appropriate to use the ideal gas law as the equation of state, a Kramers' opacity law, and a power law approximation to the energy generation from the pp chains, whereas for a much more massive star it would be appropriate to use a radiation pressure equation of state, electron scattering opacity, and a power law approximation to the energy generation from the CNO-cycles.

As an example, consider the low mass star constitutive relations for which

$$p \propto \rho T,$$
$$\kappa \propto \rho T^{-7/2}, \tag{13.15}$$
$$\varepsilon \propto \rho T^4.$$

Using (13.9) and the expression for $T$ in (13.6), these give

$$p \propto M^{a_T-3a_r+1},$$
$$\kappa \propto M^{-7a_T/2-3a_r+1}, \tag{13.16}$$
$$\varepsilon \propto M^{4a_T-3a_r+1}.$$

Comparing with equations (13.13), (13.14), and the expression for $p$ in equation (13.6), we obtain

$$a_p = a_T - 3a_r + 1,$$
$$4a_r + 4a_T - a_L - 1 = -7a_T/2 - 3a_r + 1, \tag{13.17}$$
$$a_L - 1 = 4a_T - 3a_r + 1.$$

Together with equation (13.12), we obtain a set of four simultaneous linear equations for the four homology exponents:

$$4a_r + a_p = 2,$$
$$3a_r + a_p - a_T = 1,$$
$$14a_r + 15a_T - 2a_L = 4, \tag{13.18}$$
$$3a_r - 4a_T + a_L = 2.$$

These have solution

$$a_r = 1/13, \quad a_p = 22/13, \quad a_L = 71/13, \quad a_T = 12/13. \tag{13.19}$$

Note that for the effective temperature the scaling is given by

$$T_{\text{eff}} \propto \left(\frac{L_*}{R^2}\right)^{1/4} \propto M^{a_L/4 - a_r/2}, \tag{13.20}$$

and not $a_T$. For the above example, we find $T_{\text{eff}} \propto M^{5/4}$.

## 13.3 Sensitivity of stellar structure to nuclear reaction rate

We can use homology arguments to see how sensitive global properties of the star, such as its luminosity and radius, are to the energy generation rate. We can approximate the energy generation rate from hydrogen burning by

$$\varepsilon = \varepsilon_0 \rho T^\eta, \tag{13.21}$$

where $\varepsilon_0$ and $\eta$ are constants. We can combine the electron scattering and Kramers' opacity laws into a single expression by writing

$$\kappa = \kappa_0(\alpha)\left(\rho T^{-7/2}\right)^\alpha, \tag{13.22}$$

where $\alpha = 0$ for electron scattering and $\alpha = 1$ for Kramers' opacity. The four equations for the homology exponents are now (assuming an ideal gas equation of state)

$$\begin{aligned} 4a_r + a_p &= 2 \\ 3a_r + a_p - a_T &= 1 \\ (8 + 6\alpha)a_r + (8 + 7\alpha)a_T - 2a_L &= 2(\alpha + 1) \\ 3a_r - \eta a_T + a_L &= 2. \end{aligned} \tag{13.23}$$

These have solution

$$\begin{aligned} a_r &= \frac{2\eta - 5\alpha - 2}{2\eta - \alpha + 6}, \\ a_p &= \frac{-4\eta + 18\alpha + 20}{2\eta - \alpha + 6}, \\ a_L &= \frac{(6 + 4\alpha)\eta + 13\alpha + 18}{2\eta - \alpha + 6}, \\ a_T &= \frac{4\alpha + 8}{2\eta - \alpha + 6}. \end{aligned} \tag{13.24}$$

We can write the exponent for the luminosity variable as

$$a_L = \frac{(6 + 4\alpha)\eta + 13\alpha + 18}{2\eta - \alpha + 6} = 3 + 2\alpha + \frac{2\alpha(\alpha + 2)}{2\eta + 6 - \alpha}. \tag{13.25}$$

We see that for electron scattering opacity $a_L = 3$, independent of the value of $\eta$. For Kramers' opacity

$$a_L = 5 + \frac{6}{2\eta + 5}, \tag{13.26}$$

which only weakly depends on $\eta$. Hence an important result from homology arguments is that the MS mass–luminosity relation is relatively insensitive to the temperature dependence of the nuclear energy generation rate.

## 13.4 Sensitivity of stellar properties to composition

We can use a technique similar to homology analysis to estimate how the stellar properties depend on composition. To illustrate the method, again consider constitutive relations relevant to low mass stars. The pressure is given by the ideal gas law which, for small enough heavy-element abundance and complete ionization, is

$$p = \frac{5X + 3}{4}\Re\rho T. \tag{13.27}$$

The energy generation rate due to the pp chains is approximated by

$$\varepsilon = \varepsilon_0 X^2 \rho T^4. \tag{13.28}$$

The opacity is a little problematic because in the hot interior free–free opacity will be more important than bound–free opacity, whereas in the cooler outer layers the bound–free opacity will be larger. Although both opacities have a Kramers' dependence on density and temperature, the dependence on composition is not the same. Since the heavy elements dominate the bound–free opacity and also contribute significantly to the free–free opacity, we will take

$$\kappa = \kappa_0 Z(1 + X)\rho T^{-7/2}. \tag{13.29}$$

Because of the different dependences on $X$ and $Z$, we look for solutions of form

$$r = r_1(X)r_2(Z)r_3(m), \tag{13.30}$$

with similar expressions for the pressure, luminosity, and temperature. (We do not need to introduce $q$, because we keep $M$ fixed in this analysis.)

We find

$$
\begin{aligned}
r &\propto X^{4/13}(1 + X)^{2/13}(5X + 3)^{7/13}Z^{2/13}, \\
p &\propto X^{-16/13}(1 + X)^{-8/13}(5X + 3)^{-28/13}Z^{-8/13}, \\
L &\propto X^{-2/13}(1 + X)^{-14/13}(5X + 3)^{-101/13}Z^{-14/13}, \\
T &\propto X^{-4/13}(1 + X)^{-2/13}(5X + 3)^{-20/13}Z^{-2/13}.
\end{aligned}
\tag{13.31}
$$

Note that these relations predict that stars of lower $Z$ will be smaller, more luminous, and hotter at the surface than stars of higher $Z$ and the same mass (and hydrogen mass fraction). Since $Z$ only enters through the opacity, these differences must be due to the lower opacity in stars of lower $Z$. A lower opacity means it is easier for radiation to leave the star and hence the star is more luminous. To generate the greater luminosity the central temperature must be higher. A higher central temperature leads to higher central pressure. To balance a larger pressure gradient, the gravity must be higher and hence the star must be smaller.

The same changes occur if $Z$ is kept fixed and $X$ is decreased. Hence fully mixed stars would evolve to the blue during hydrogen burning. Both of the last two results are borne out by detailed calculations.

## 13.5 Stars with convective cores

Since the density is high, when convection occurs in the core of a star it is very efficient and so

$$\frac{d \ln T}{d \ln p} = \nabla_{ad}. \tag{13.32}$$

Applying the homology relations (13.6),

$$\frac{d \ln \tilde{T}}{d \ln \tilde{p}} = \nabla_{ad}, \tag{13.33}$$

independent of the values of $a_p$ and $a_T$. Hence in the core there are too few equations to determine the homology exponents. However, continuity of temperature and pressure at the boundary between the convective core and the radiative envelope indicate that there is a solution in which $a_p$ and $a_T$ are determined by conditions in the radiative envelope and, assuming the constitutive relations do not change, the convective–radiative transition occurs at a value of $q$ that is independent of $M$. Since massive stars have convective cores and radiative envelopes, these considerations show that homology arguments can be used for upper MS stars.

## 13.6 Stars with convective envelopes

Low mass stars, like the Sun, have convective envelopes. Because the convection results mainly from the high opacity associated with photo-ionization, the adiabatic gradient is not constant. Furthermore, in the outer parts of the convection zone, the density can be low enough that convection is not efficient. Hence the structural gradient can vary significantly with position in the convective envelope. This means that homology is not a good approximation for lower MS stars, as is evident from the large difference between the observationally determined mass–radius relation, $R \propto M^{0.9}$, and that obtained from homology arguments, $R \propto M^{1/13}$.

# Chapter 14

## Hydrogen main sequence stars

### 14.1 Masses of main sequence stars

The lower limit on the mass of a hydrogen burning MS star is set by the requirement that its central temperature be high enough for the p + p reaction to occur at a rate such that the star is in thermal equilibrium. For solar composition material, this lower limit is about 0.072 $M_\odot$ [1]. Stars with mass less than this limit are brown dwarf stars. It is not yet understood what sets the upper limit to the mass of MS stars. The current (as of July 2015) record holder for the highest accurate mass measured in a binary system is NGC 3603 A1 [2], which has a mass of 116 ± 31 $M_\odot$. From its luminosity of about 5–40 × $10^6$ $L_\odot$, LBV 1806-20 is estimated to have a mass of about 150 $M_\odot$. The WN star R136a1 [3] in the central cluster of the Tarantula nebula (30 Doradus) has a luminosity of 9 × $10^6$ $L_\odot$, which implies a mass >170 $M_\odot$. The initial mass of this evolved star may have been larger than 225 $M_\odot$. It is possible that more massive stars exist in our Galaxy but have not been discovered due to large amounts of interstellar extinction. Very massive stars are rare and short lived with MS lifetimes of about 3 million years. It is likely that radiation pressure plays a role in setting the upper mass limit, either by restricting accretion of material during formation of the star or by destabilizing the star. For a pure radiation equation of state (section 6.3), $U = 3\int_0^M \frac{p}{\rho}\mathrm{d}m$, and application of the virial theorem (section 3.3) shows that a star in hydrostatic equilibrium supported by pure radiation pressure has zero total energy and hence is easily disrupted.

### 14.2 Lifetimes of main sequence stars

Figure 14.1 shows as a function of mass the PMS and MS lifetimes for solar composition stellar models. The value of the mixing length ratio, $\alpha = 1.7$, used to construct these models was determined by matching models to solar data. The PMS lifetime is the time for the model to contract from a low density state to the point at which the main hydrogen burning phase begins. The MS lifetime is the time the

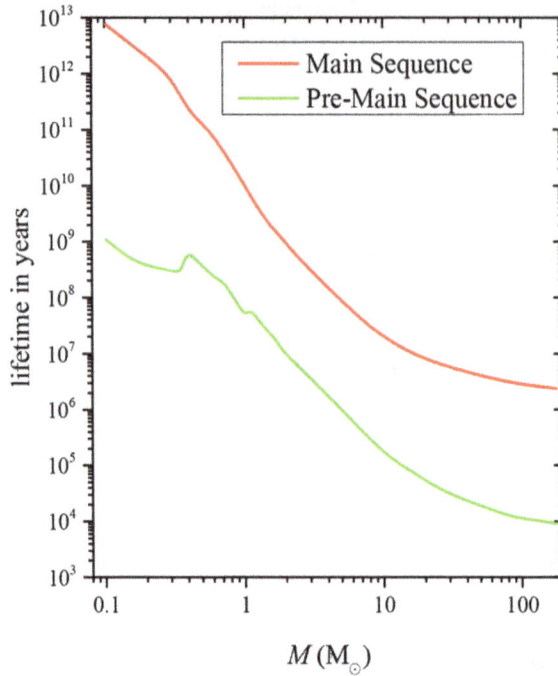

**Figure 14.1.** PMS and MS lifetimes for solar composition stellar models.

model spends in the core hydrogen burning phase, which here is taken to end at the earlier of either the gravo-thermal power exceeding 1% of the stellar luminosity or the hydrogen abundance at the stellar center decreasing to $X = 10^{-5}$. Both lifetime scales are relevant to dating clusters by fitting evolutionary models to cluster CMDs. For example, in young clusters of age a few 100 million years, the low mass stars of mass less than about 0.6 $M_\odot$ will still be contracting to the MS and these stars will appear above the MS. Given that the age of the Universe is about 14 billion years, we see that stars of mass less than 0.95 $M_\odot$ will not leave the MS. We also see that the most massive stars have very short lifetimes, of a few million years, and will leave the MS before stars of mass less than 3 $M_\odot$ have reached the MS.

## 14.3 Convection in main sequence stars

Models of ZAMS stars show that solar composition stars of mass less than about 0.35 $M_\odot$ are fully convective. Stars with mass greater than 1.3 $M_\odot$ are convective in their cores. The convective regions of MS stars are shown in gray in figure 14.2. The left-hand panel has a linear scale for the ordinate and the right-hand panel has a logarithmic scale. The right-hand panel reveals the intricate structure of the surface convection zones in massive stars. Note that these convection zones contain relatively little mass. The labels indicate the reason for convective instability, with H, He, and He$^+$ indicating ionization of hydrogen, neutral helium, and singly ionized helium, respectively.

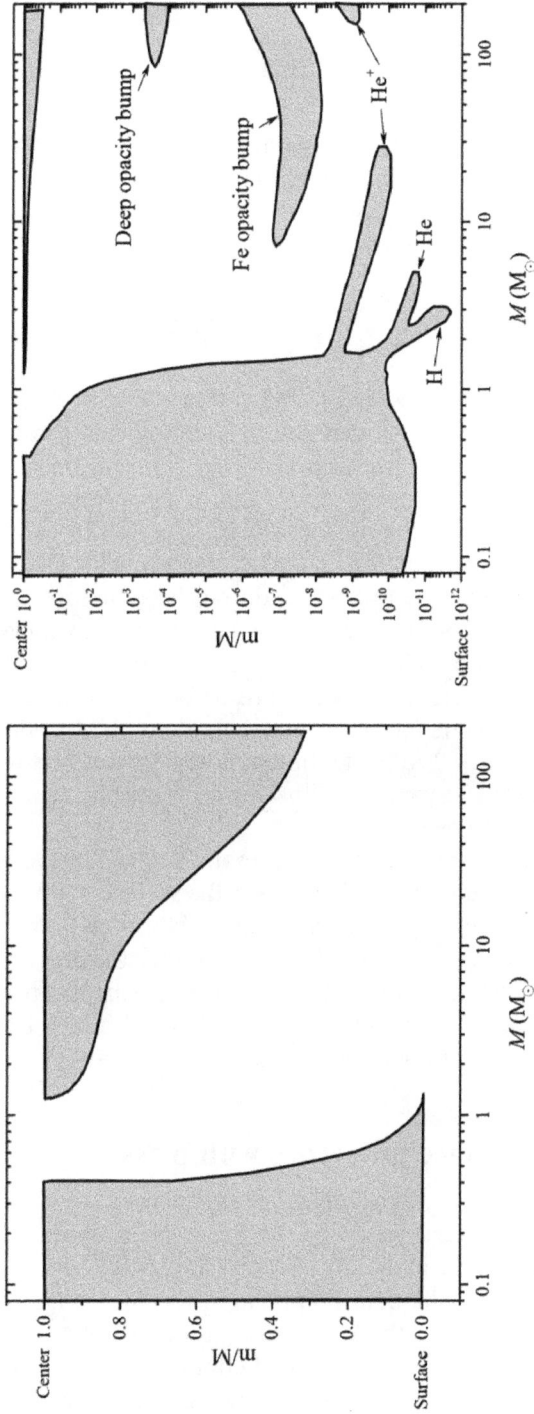

**Figure 14.2.** Location of the convection zones of MS stars.

To understand why massive stars have convective cores, consider the expression for the radiative gradient

$$\nabla_{\text{rad}} = \frac{1}{4} \frac{3p}{aT^4} \frac{\kappa L}{4\pi c Gm}.$$  (14.1)

If the nuclear reactions are concentrated in a region of small mass at the center of the star, then $L/m$ will be very large in this region, so that $\nabla_{\text{rad}}$ is also large there and, according to the Schwarzschild criterion, the material will be convective. For massive stars, the energy is generated by the CNO-cycle for which

$$\frac{\partial \ln \varepsilon_{\text{CNO}}}{\partial \ln T_6} = -\frac{2}{3} + \frac{50.77}{T_6^{1/3}}.$$  (14.2)

Hence at the central temperatures that occur in massive stars, $\varepsilon_{\text{CNO}} \propto T_6^{17}$. It is this steep dependence of the energy generation rate on temperature that makes the energy generating region of the star centrally concentrated and consequently the inner parts of the star are convective. For less massive stars, such as the Sun, the energy is generated by the pp chains, for which $\varepsilon_{\text{pp}} \propto T_6^4$. In this case, the energy generating region has a much larger extent and $L/m$ is much smaller, and the core is radiative (provided the other factors in $\nabla_{\text{rad}}$ do not become too large.)

The massive stars do not have extensive convective envelopes because their surface temperatures are too high for there to be a hydrogen partial ionization zone. The thin convection zones that do occur are associated with the helium partial ionization zones and, for stars more massive than about 8 $M_\odot$, with the Fe opacity 'bump' that occurs near $T \sim 2 \times 10^5$ K. In the most massive stars with $M > 85\ M_\odot$, there is a convection zone associated with the 'deep opacity bump' due mainly to ionization of Fe at $T \approx 1.7 \times 10^6$ K.

Stars with surface temperatures less than about 10 000 K have hydrogen partial ionization zones in their outer layers. The lower the surface temperature, the deeper the ionization zone. Also in the cooler stars, dissociation of $H_2$ molecules becomes an important source of opacity in the outer layers. Hence the depth of the convection zone is larger for cooler stars. Note that the radiative gradient is proportional to the opacity and also to $p/p_{\text{rad}}$. In low mass stars, both these factors become significant in the central regions, so that stars with mass less than about 0.35 $M_\odot$ are fully convective.

## 14.4 Variation of surface properties with mass

Figure 14.3 shows how the range of radii experienced by models during MS evolutions depends on stellar mass. For the lower mass models the evolutionary endpoint has been taken to be at age $10^{10}$ years. Also shown are the binary star data of Torres *et al* [4]. Most of the data points lie in the MS range. A few stars with $M > 1\ M_\odot$ have evolved beyond the MS to larger radii. A noticeable problem occurs for the lowest mass stars which have not had enough time to evolve from the MS. These stars are larger and, as we will see later, cooler than the models predict. A probable resolution is that these stars, which are in tidally locked short period

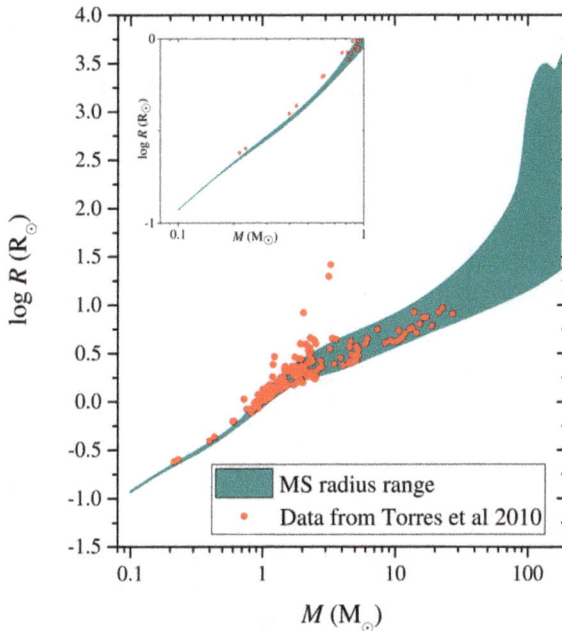

**Figure 14.3.** MS radius range for models of solar composition.

binary systems, have magnetic fields strong enough to inhibit convective energy transport, which results in larger and cooler models [5, 6].

The most massive stars shown here experience large radius increases during MS evolution. However there is observationally evidence that the expansion is not as large as shown here [7]. The probable explanation is that mass loss from stellar winds or eruptions significantly reduces the stellar mass during MS evolution.

Note the change in slope of the radius–mass relation near 1.5 $M_\odot$, due to the transition from lower mass stars with radiative cores and convective envelopes to higher mass stars with convective cores and radiative envelopes.

Figures 14.4 and 14.5 are similar to figure 14.3, except they show how the MS luminosity and effective temperature ranges depend on mass.

Again we see that most of the binary data points lie on the MS, with the most notable exception being the lowest mass stars, which are cooler than the models predict.

We also see that stars more massive than about 2.5 $M_\odot$ have $T_{\text{eff}} > 10\,000$ K and hence these stars are indeed too hot to have a hydrogen partial ionization zone. Also note that there is a maximum to $T_{\text{eff}}$ of about 50 000 K.

## 14.5 Variation of central properties with mass

Figures 14.6 and 14.7 show, respectively, the range of temperature and density experienced at the model center during MS evolution as a function of stellar mass.

We see from the lower boundary in figure 14.6 that the transition in energy generation from the pp-chains to the CNO-cycles at $T_c = 18 \times 10^6$ K occurs at about

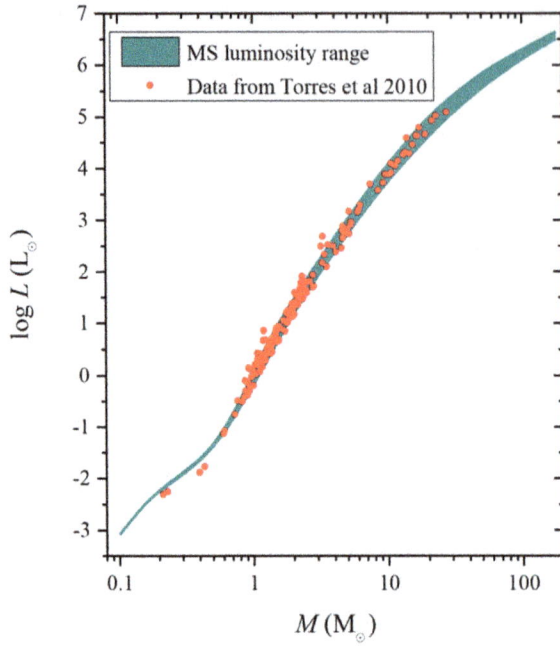

**Figure 14.4.** MS luminosity range for models of solar composition.

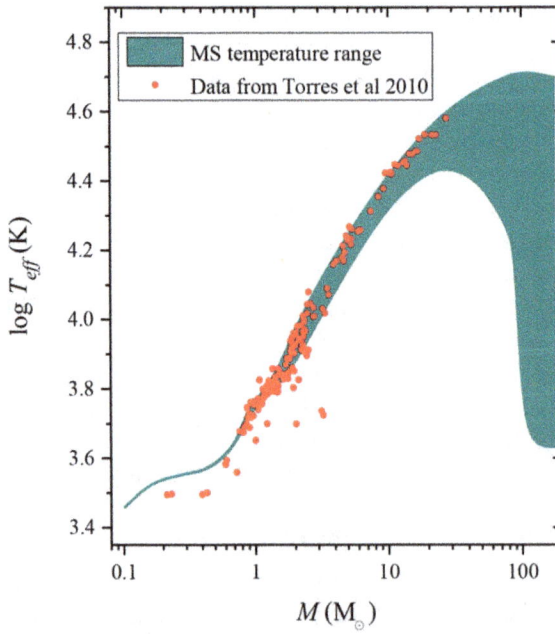

**Figure 14.5.** MS temperature range for models of solar composition.

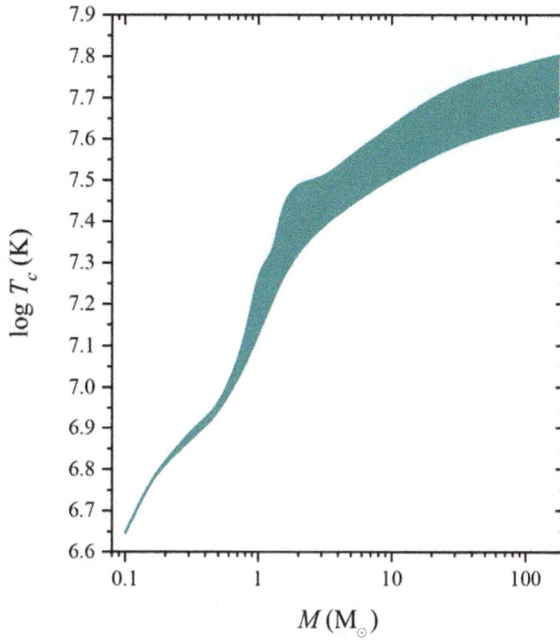

**Figure 14.6.** MS central temperature range for solar composition models.

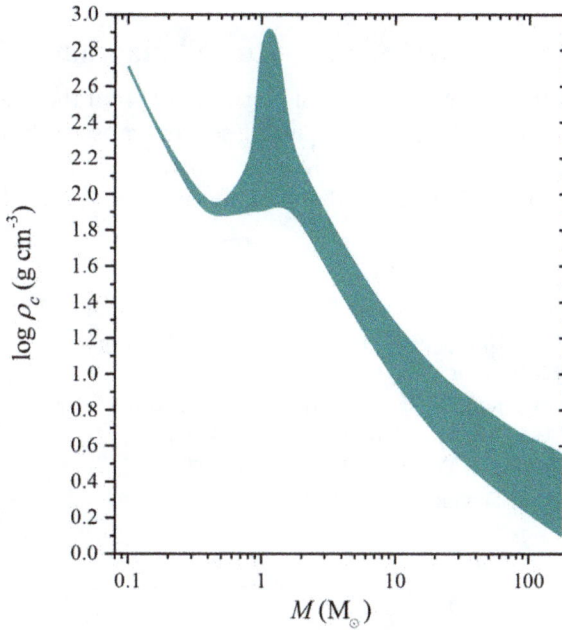

**Figure 14.7.** MS central density range for solar composition models.

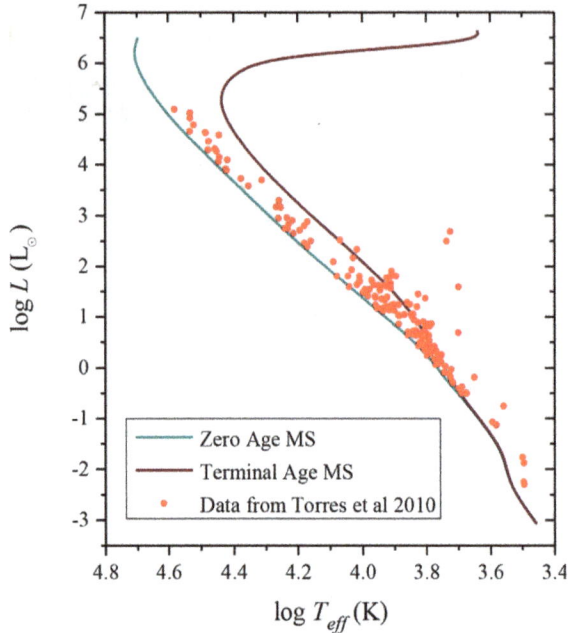

**Figure 14.8.** Theoretical HRD for solar composition stellar models. Data from [4].

1.5 $M_\odot$ for ZAMS stars. From figure 14.7, we see that for stars which are convective at the center, the density decreases as mass increases but for stars with radiative cores, the density increases with mass.

## 14.6 The theoretical Hertzsprung–Russell diagram

Figure 14.8 shows the binary star data of Torres *et al* [4] in the HRD together with the ZAMS and the terminal age MS. Again we see that most of the data points lie on the MS. A few stars with $T_{eff} > 4000$ K have evolved off the MS. As described in section 14.3, we also see that the coolest stars do not lie on the MS.

## Bibliography

[1] Chabrier G and Baraffe I 2000 *Annu. Rev. Astron. Astrophys.* **38** 337
[2] Schnurr O *et al* 2008 *Mon. Not. R. Astron. Soc.: Lett.* **389** L38
[3] Crowther P *et al* 2010 *Mon. Not. R. Astron. Soc.* **408** 731
[4] Torres G, Andersen J and Giménez A 2010 *Astron. Astrophys. Rev.* **18** 67
[5] Mullan D J and MacDonald J 2001 *Astrophys. J.* **559** 353
[6] MacDonald J and Mullan D J 2014 *Astrophys. J.* **787** 70
[7] Humphreys R M and Davidson K 1979 *Astrophys. J.* **232** 409

# Chapter 15

## Helium main sequence stars

### 15.1 Why consider helium main sequence stars?

A helium MS star could be formed by the evolution of a fully mixed star to hydrogen exhaustion. However cluster CMDs indicate that the majority of stars evolve to the red during core hydrogen burning and hence by homology arguments they are not fully mixed. Since massive stars have convective cores, their central regions are, however, fully mixed. A helium star could also be formed if the hydrogen-rich envelope were removed by mass loss, e.g. in a close binary system. After the end of core hydrogen burning in a low mass star, a helium core builds up due to shell hydrogen burning. As the helium core grows in mass, it contracts and heats. When the core temperature is high enough, the $3\alpha$ process begins and the star will eventually settle into a core helium burning phase. Again, if the hydrogen-rich outer layers of the star are lost by some mechanism, a helium MS star is formed. Even in the case that not all of the hydrogen-rich layer is lost, models of the helium MS can place limits on the effective temperature of the star.

Models of helium stars have proven useful in the study of the evolution of massive stars through the stages leading up to core-collapse supernovae [1]. The lower limit on the mass of a helium burning MS star is set by the requirement that its central temperature be high enough for the $3\alpha$ reaction to occur at a rate that the star is in thermal equilibrium. For solar heavy-element abundances, this lower limit is about $0.35\ M_\odot$. The upper limit to the mass of a helium MS star is set by the mass of the convective core of its progenitor. At the end of core hydrogen burning, a $100\ M_\odot$ star has a helium core of about $45\ M_\odot$.

### 15.2 Homology analysis of helium zero age main sequence stars

We first consider the case in which the opacity is constant. This is appropriate for the hot interiors of helium stars in which electron scattering is the main opacity source. We will also assume that the pressure is given by the ideal gas law

$$p = \frac{\mathfrak{R}}{\mu}\rho T, \tag{15.1}$$

and that the energy generation rate is of form

$$\varepsilon = \varepsilon_0 \rho^\lambda T^\eta. \tag{15.2}$$

The equations for the homology exponents are

$$
\begin{aligned}
&4a_r + a_p - 2 = 0, \\
&a_p = a_T - 3a_r + 1, \\
&4a_r + 4a_T - a_L - 1 = 0, \\
&a_L - 1 = \eta a_T - 3\lambda a_r + \lambda.
\end{aligned}
\tag{15.3}
$$

Eliminating $a_p$ from the first two equations gives

$$a_r + a_T = 1, \tag{15.4}$$

which when inserted into the third equation gives

$$a_L = 3. \tag{15.5}$$

Hence the mass–luminosity relation is independent of $\lambda$ and $\eta$, and hence by extension to $\varepsilon_0$. Keeping track of the dependence on $\mu$ and $\kappa$, we obtain

$$L \propto \frac{\mu^4 M^3}{\kappa}. \tag{15.6}$$

Hence, assuming the same opacity and pressure relations, for stars of the same mass, the He ZAMS is more luminous than the H ZAMS by a factor

$$\frac{L_{\text{He}}}{L_{\text{H}}} = \left(\frac{\mu_{\text{He}}}{\mu_{\text{H}}}\right)^4 \frac{\kappa_{\text{H}}}{\kappa_{\text{He}}} = \left(\frac{6.5}{3}\right)^4 1.7 = 37.5. \tag{15.7}$$

With $\lambda = 2$ and $\eta = 40$, appropriate values for the $3\alpha$ reaction, we obtain

$$a_r = \frac{20}{23}, \quad a_p = -\frac{34}{23}, \quad a_L = 3, \quad a_T = \frac{3}{23}. \tag{15.8}$$

Again, keeping track of $\mu$, $\kappa$ and also $\varepsilon_0$, we find

$$R \propto (\kappa \varepsilon_0)^{1/46} \mu^{18/23} M^{20/23}, \tag{15.9}$$

so that

$$T_{\text{eff}} \propto \kappa^{-1/4}(\kappa \varepsilon_0)^{-1/92} \mu^{14/23} M^{29/92}. \tag{15.10}$$

Again we see that the radius and effective temperature are only weakly dependent on $\varepsilon_0$, indicating that it is mainly the stellar structure that sets the luminosity and not the energy generation rate.

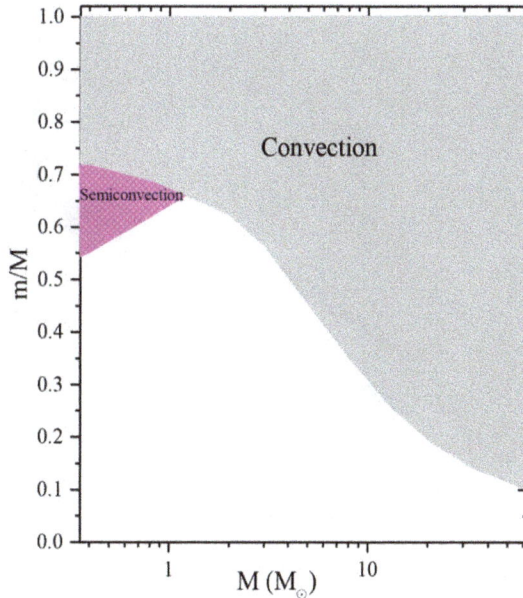

**Figure 15.1.** Location of convection zones in He MS stars.

## 15.3 Convection in helium main sequence stars

Because the 3α reaction is very sensitive to temperature ($\varepsilon_{3\alpha} \propto T^{40}$), He MS stars have convective cores for the same reason that massive H MS stars have convective cores. Also, because He MS stars have high effective temperatures and high gravities, surface convection zones are thin. Figure 15.1 shows the location of the convection zones for models of He MS stars at the midpoint of their MS evolution. The region labelled semiconvection shows where the convection criterion is affected by the presence of composition gradients.

## 15.4 Variation of surface properties with mass

Figure 15.2 compares the ranges of radii of helium MS stars with those of hydrogen MS stars. For the helium MS stars there is a noticeable change in slope near 10 $M_\odot$ due to the increasing importance of radiation pressure with mass. For masses less than about 10 $M_\odot$, the He MS stars are smaller than the H MS stars by a factor of about 5. This is due to the higher temperature required for helium burning reactions compared to hydrogen burning reactions. If we crudely assume that the He ZAMS and H ZAMS stars are homologous, and also that the ideal gas law for pressure is applicable (which it is at the lower mass parts of the sequences), then the scaling of central temperature with stellar mass, stellar radius and molecular weight is

$$T \propto \frac{\mu M}{R}.$$ (15.11)

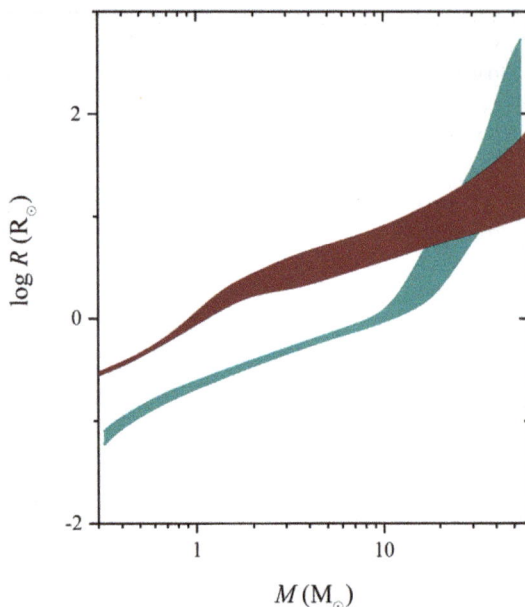

**Figure 15.2.** Comparison of the ranges of radii of He and H MS stars.

Hence at a given mass the ratio of radii is

$$\frac{R_H}{R_{He}} = \frac{\mu_H}{\mu_{He}} \frac{T_{He}}{T_H} \approx 5, \qquad (15.12)$$

since $T_{He}/T_H \approx 10$.

Figure 15.3 compares the luminosity ranges of helium and hydrogen MS stars. As indicated by equation (15.7), the He MS stars are more luminous than the H MS stars of the same mass. Figure 15.4 compares the model He MS range of $T_{eff}$ with the model H MS range. For stars with $M < 10 M_\odot$, the He stars have surface temperatures that are about a factor 5 higher than those of H stars of the same mass.

Also the lower mass He MS stars have effective temperatures greater than 30 000 K and hence are too hot to have He–He$^+$ partial ionization zones. Also note that there is a maximum to the temperature of about 120 000 K at mass $\approx 15 M_\odot$. This is due to the increasing importance of radiation pressure in the more massive stars.

## 15.5 Variation of central properties with mass

Figure 15.5 compares the central temperature ranges of He MS and H MS stellar models.

We see that because temperatures in excess of $10^8$ K are needed for helium burning, the central temperature of a He MS star is significantly higher than that of a H MS star of the same mass.

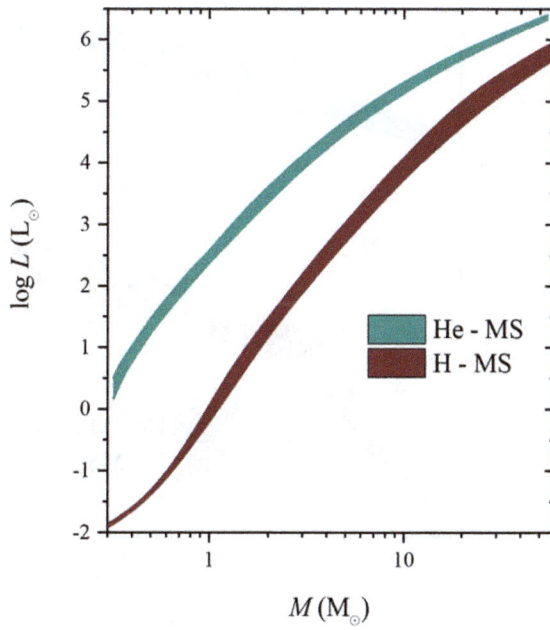

**Figure 15.3.** Comparison of the ranges of luminosities of He and H MS stars.

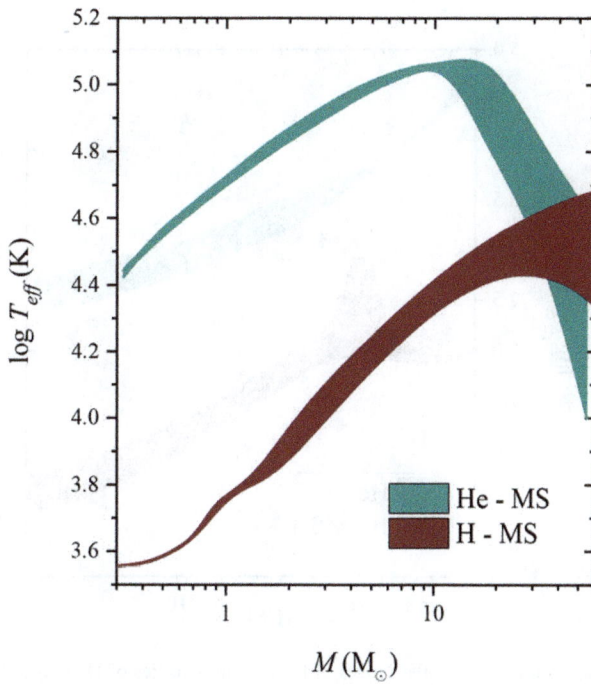

**Figure 15.4.** Comparison of the ranges of effective temperatures of He and H MS stars.

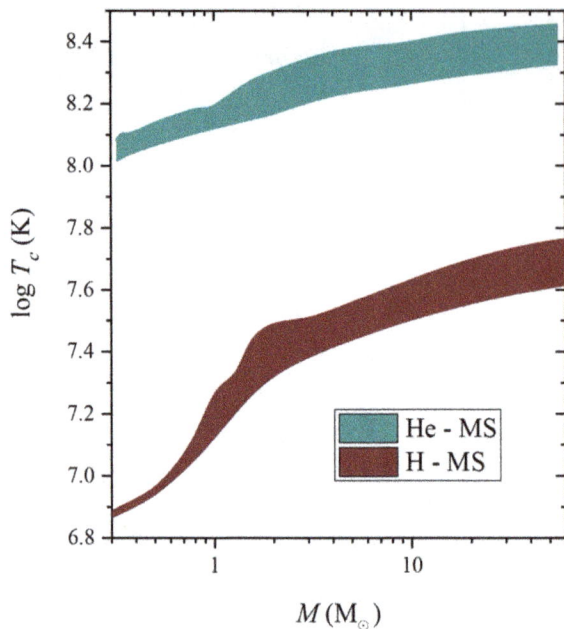

**Figure 15.5.** Comparison of the ranges of the central temperatures of He and H MS stars.

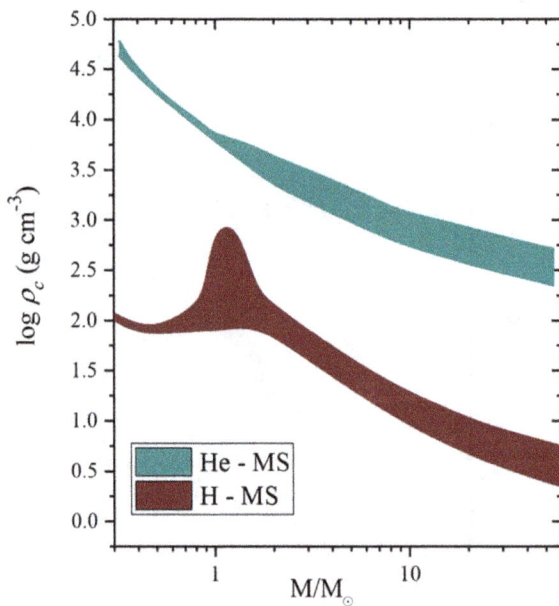

**Figure 15.6.** Comparison of the ranges of the central densities of He and H MS stars.

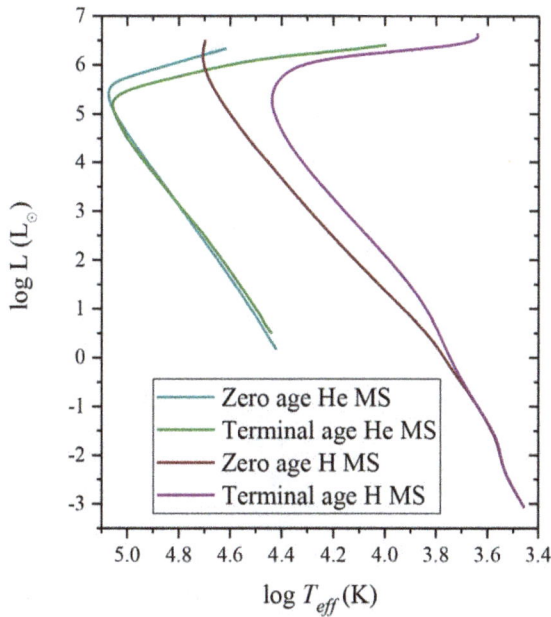

**Figure 15.7.** Comparison of He MS and H MS stars in the HRD.

Figure 15.6 compares the central density ranges of He and H MS models. Since the He MS stars are convective at the center, their central density decreases as mass increases.

## 15.6 The theoretical Hertzsprung–Russell diagram

In figure 15.7 we compare the HRD for model He MS stars with that for H MS. We see that, except for the most luminous stars, the He MS is bluer than the H MS.

## Bibliography

[1] Woosley S E, Heger A and Weaver T A 2002 *Rev. Mod. Phys.* **74** 1015

Structure and Evolution of Single Stars
An introduction
**James MacDonald**

# Chapter 16

## The Hayashi line

## 16.1 Introduction

Here we begin investigation of evolutionary phases off the MS by considering the Hayashi line, which is relevant to evolution on the RGB and also to part of the early phases of evolution before the MS is reached. In CMDs of old clusters, we find that the RGB spans a range in luminosity of more than 100 yet the stellar temperature has a relatively small range from about 5000–3000 K. Indeed there are very few stars (excluding the brown dwarfs) that have temperature less than about 3000 K. In 1961, Hayashi [1] showed that in models of fully convective stars there is a dividing line between an 'allowed' region of the HRD and a 'forbidden' region. This dividing line is now called the Hayashi line. It approximately describes the path in the HRD taken by stars evolving on the RGB. Figure 16.1 shows the evolutionary track of a 1 $M_\odot$ model star of solar composition from the PMS phase through core hydrogen burning and on to the beginning of helium burning in the core.

During the PMS Hayashi phase the star is completely convective except for a small outer part, and on the RGB the star has an extensive convection zone in its outer layers. To obtain some insight into why there are extensive convection zones during these phases, consider again the expression for the radiative gradient

$$\nabla_{\mathrm{rad}} = \frac{1}{4}\frac{3p}{aT^4}\frac{\kappa L}{4\pi cGm} = \frac{1}{4}\frac{p}{p_{\mathrm{rad}}}\frac{\kappa L}{4\pi cGm}. \tag{16.1}$$

In the outermost layers of the star, because of the low temperature, the hydrogen is neutral and the opacity is low. These outer parts are radiative. As we go deeper into the star, the temperature increases and hydrogen starts to become ionized. In the partial ionization zone, the opacity increases to large values and the adiabatic gradient decreases. These layers are convective. Provided the density is high enough, convection will be efficient and the structural gradient is equal to the adiabatic gradient. Because the adiabatic gradient is small (as low as 0.1), the temperature

doi:10.1088/978-1-6817-4105-5ch16
16-1

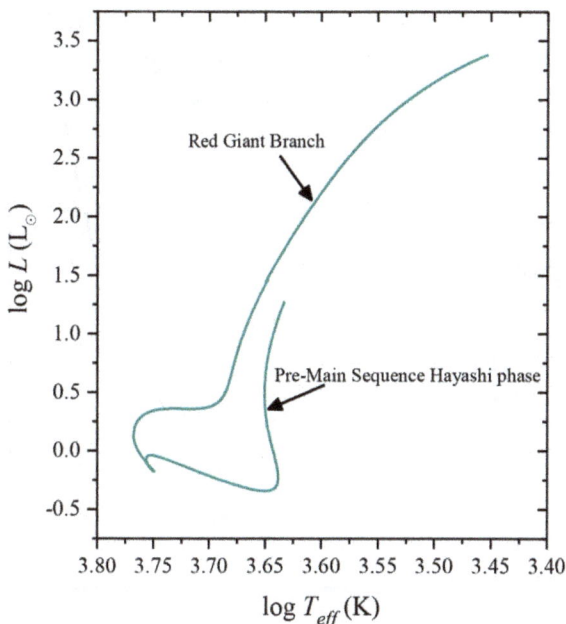

**Figure 16.1.** Evolutionary path of a 1 $M_\odot$ model star in the HRD from the PMS to the red giant tip.

does not increase as fast as the pressure as we go deeper into the star. As a consequence the factor $p/p_{rad}$ becomes large so that when the opacity decreases due to ionization becoming complete, $\nabla_{rad}$ remains very large.

The envelope remains convective until depths are reached such that $p/p_{rad}$ is reduced enough that the envelope can become radiative again. This behavior is illustrated in figure 16.2, where the factor $p/p_{rad}$ is shown against $p$ for a PMS star model and for an RGB model. The red star is placed at the location of the bottom of the surface convection zone of the red giant star. For the PMS Hayashi phase, the star remains convective down to its center.

## 16.2 The Hayashi phase

The Hayashi phase is a possible evolutionary path taken by a star in its earliest stages after formation. Here we use a simple model to derive an approximate relation between the luminosity and effective temperature that shows how the star evolves on the HRD during this phase. Because of the similar structure in red giant stars, we can also apply it to these objects. The simple model involves matching a radiative solution for the stellar structure above the photosphere to a convective solution for the structure below the photosphere. We ignore complexities that arise from inefficient convection and non-uniformity of the adiabatic gradient.

In the Hayashi phase the star is large and has a surface temperature low enough that the dominant constituent, hydrogen, is neutral. In the outer radiative parts of

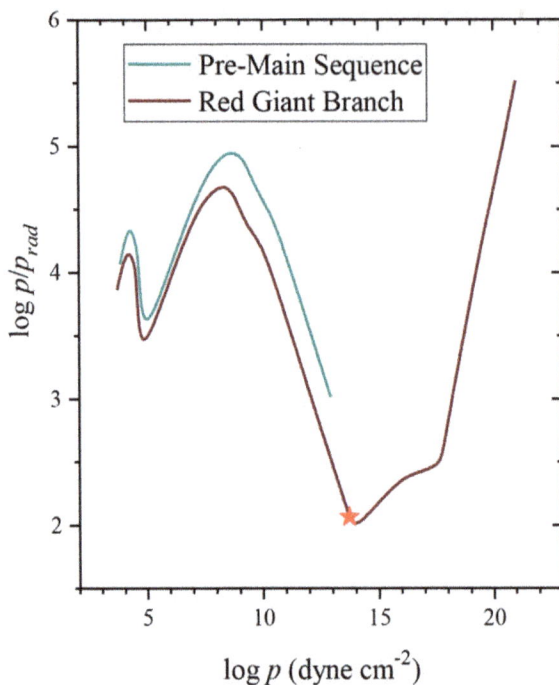

**Figure 16.2.** Ratio $p/p_{\text{rad}}$ plotted against pressure for PMS and red giant models.

the star, the opacity is low and increases with temperature. We can approximate the opacity in these regions by

$$\kappa = \kappa_0 \rho^\alpha T^\beta, \tag{16.2}$$

where $\alpha$ and $\beta$ are positive.

At the photosphere, we have the conditions

$$p\kappa = \frac{2}{3}g, \tag{16.3}$$

and

$$L = \pi a c R^2 T^4. \tag{16.4}$$

For an ideal gas equation of state, the first photospheric condition gives

$$\frac{\Re}{\mu} \kappa_0 \rho^{\alpha+1} T^{\beta+1} = \frac{2}{3} \frac{GM}{R^2}. \tag{16.5}$$

To obtain the relation between the photospheric temperature and the luminosity, we need an expression for the density at the photosphere. If we assume that (i) below the photosphere the star is completely convective, (ii) convection is efficient, and (iii) the adiabatic gradient is that for an ideal gas in the absence of ionization, we have

$$p = K\rho^{5/3}, \tag{16.6}$$

16-3

where $K$ is a constant that depends on the properties of the star. Hence the star is a polytrope of index $n = 3/2$. The radius and mass of the star in this polytropic model are given by

$$R = l\xi_1,$$

(16.7)

and

$$M = 4\pi l^3 \rho_c \left( -\xi^2 \frac{d\theta}{d\xi} \right)\bigg|_{\xi=\xi_1},$$

(16.8)

where $\rho_c$ is the central density and the length scale, $l$, is

$$l = \sqrt{\frac{5K}{8\pi G \rho_c^{1/3}}}.$$

(16.9)

By eliminating $l$ and $\rho_c$, we obtain an expression for $K$ in terms of $M$ and $R$:

$$K = \left[ \frac{R^2}{\xi_1^2} \frac{8\pi G}{5} \right] M^{1/3} \left[ 4\pi R^3 \left( -\frac{1}{\xi} \frac{d\theta}{d\xi} \right)\bigg|_{\xi=\xi_1} \right]^{-1/3} = 5.751 \times 10^3 \left( \frac{M}{M_\odot} \right)^{1/3} R.$$

(16.10)

(Note cgs units are being used.) Using the ideal gas law, we also have

$$K = \frac{\mathfrak{R}}{\mu} \frac{T}{\rho^{2/3}}.$$

(16.11)

Eliminating $K$ from the last two equations, we find

$$\rho = \rho_0 \left( \frac{\mathfrak{R}}{\mu} \frac{T}{R} \right)^{3/2} \left( \frac{M}{M_\odot} \right)^{-1/2},$$

(16.12)

where $\rho_0 = 2.293 \times 10^{-6}$ in cgs units. From equation (16.5), we find that at the photosphere the temperature is given by

$$\left( \frac{\mathfrak{R}}{\mu} \right)^{(3\alpha+5)/2} \kappa_0 \rho_0^{\alpha+1} T^{(3\alpha+2\beta+5)/2} = \frac{2}{3} G M_\odot \left( \frac{M}{M_\odot} \right)^{(\alpha+3)/2} R^{(3\alpha-1)/2}.$$

(16.13)

Eliminating $R$ from equations (16.4) and (16.13), we finally obtain that at the photosphere

$$\left( \frac{\mathfrak{R}}{\mu} \right)^{(3\alpha+5)} \kappa_0^2 \rho_0^{2\alpha+2} T^{9\alpha+2\beta+3} = \frac{4}{9} (G M_\odot)^2 \left( \frac{M}{M_\odot} \right)^{\alpha+3} (2.324 \times 10^{18})^{3\alpha-1} \left( \frac{L}{L_\odot} \right)^{(3\alpha-1)/2}.$$

(16.14)

To find the evolutionary path in the HRD, we need appropriate values for $\kappa_0$, $\alpha$, and $\beta$. Figure 16.3 is an opacity contour plot. The contours are labeled with the log of the opacity in units of $cm^2\ g^{-1}$. The run of density and temperature for a Hayashi phase model is also shown, with the star marking the location of the photosphere. Figures 16.4 and 16.5 show contours of the exponents $\alpha$ and $\beta$.

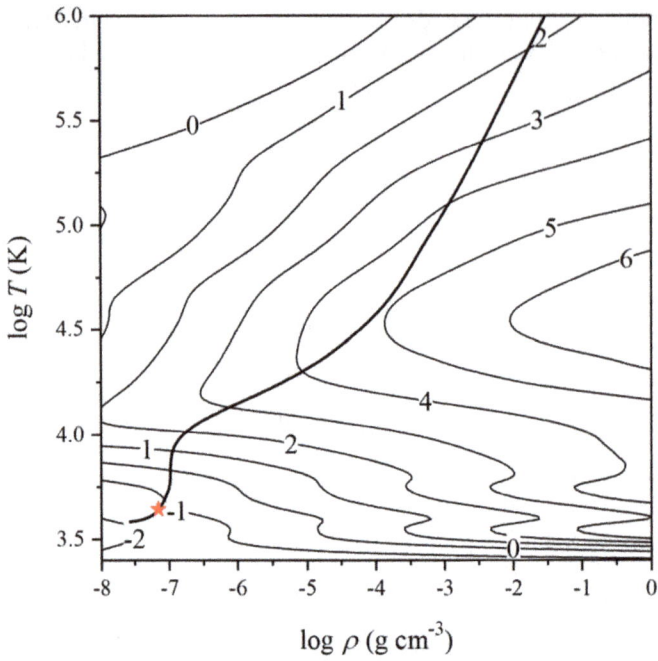

**Figure 16.3.** Contours of radiative opacity in the $\log\rho$–$\log T$ plane.

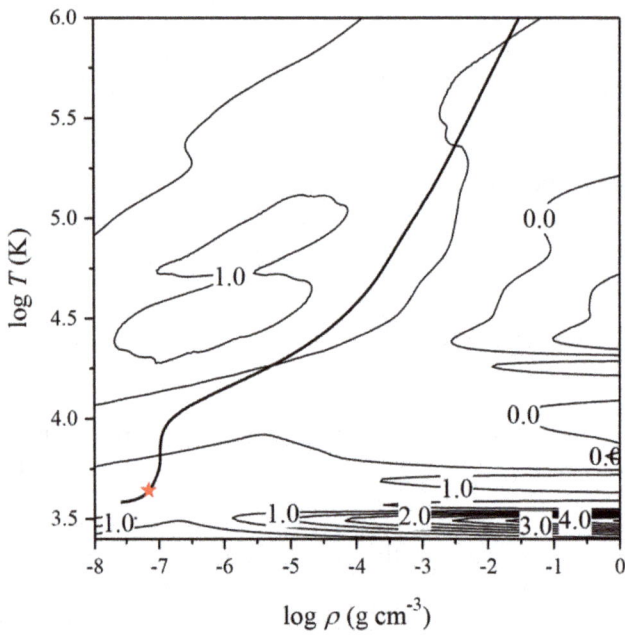

**Figure 16.4.** Contours of $\alpha = \partial \log \kappa / \partial \log \rho$.

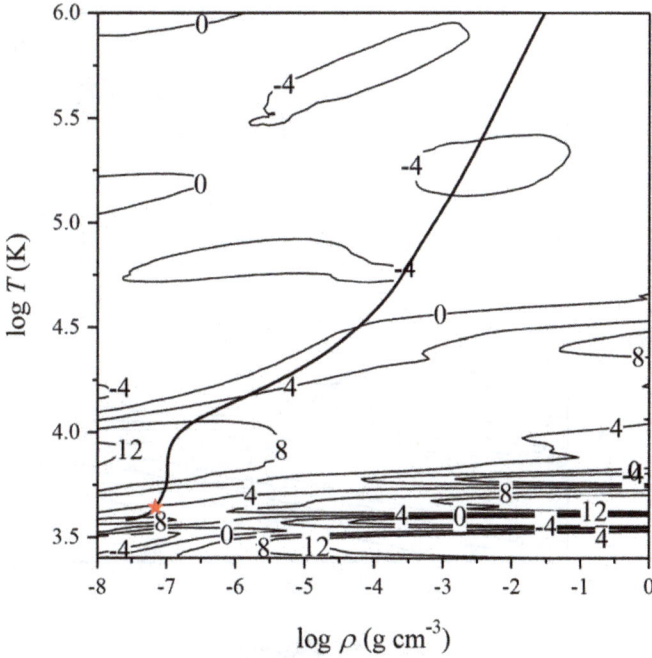

**Figure 16.5.** Contours of $\beta = \partial \log \kappa / \partial \log T$.

Since the contours in the figure on the right are close together at the bottom, an enlargement is given in figure 16.6.

We see that near the photosphere $\beta$ ranges from 2–8 and $\alpha \simeq 0.8$. Inspection of the exponents in equation (16.14) shows that the temperature will depend only weakly on luminosity. Taking $\beta = 6$ and fitting to the opacity value at the photosphere, we find

$$\kappa \simeq 5.5 \times 10^{-18} \rho^{0.8} T^6 \text{ cm}^2 \text{ g}^{-1}. \tag{16.15}$$

Using this in equation (16.14), we obtain for the photospheric temperature

$$T = 2.30 \times 10^3 \left(\frac{M}{M_\odot}\right)^{0.17} \left(\frac{L}{L_\odot}\right)^{0.032} \text{ K}. \tag{16.16}$$

This confirms that the effective temperature is very weakly dependent on the luminosity, and hence this simple model indicates that the star should evolve vertically in the HRD, which is in agreement with what is found from detailed models. Also equation (16.16) indicates that the effective temperature does not depend very strongly on the mass, which is also found in the detailed models.

For densities and temperatures appropriate to the atmospheres of RGB stars, $\alpha \simeq 0.65$, $\beta \simeq 8$, and $\kappa_0 \simeq 2.7 \times 10^{-26}$. In this case

$$T = 7.56 \times 10^2 \left(\frac{M}{M_\odot}\right)^{0.147} \left(\frac{L}{L_\odot}\right)^{0.019} \text{ K}. \tag{16.17}$$

Again the evolution is predicted to occur at an almost constant temperature.

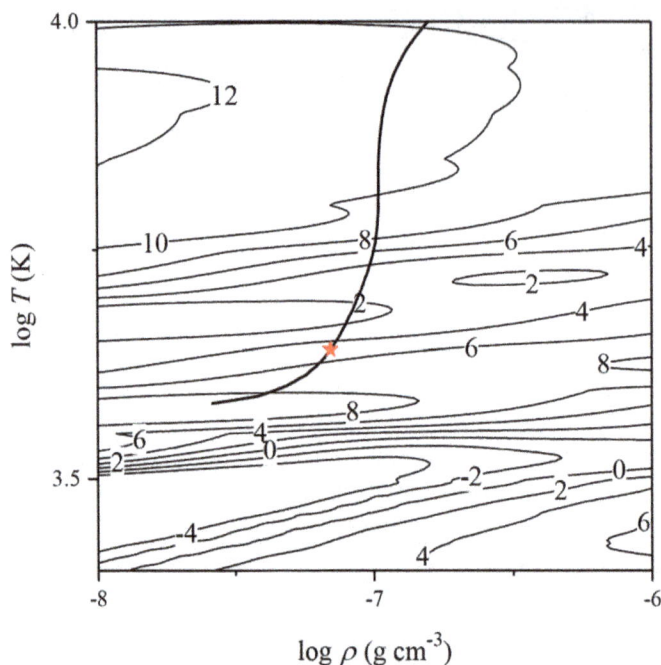

**Figure 16.6.** An enlargement of the lower part of figure 16.5.

The existence of the Hayashi line is essentially an optical depth effect. Suppose the temperature at the photosphere were increased by a small amount. Because of the strong dependence of the opacity on temperature, the relative opacity increase will be much larger, which significantly increases the optical depth to the 'photosphere'. In consequence, the photosphere would move further out to regions of lower opacity and temperature. Likewise, if the photospheric temperature were decreased, the photosphere would move into regions of higher opacity and temperature.

## Bibliography

[1] Hayashi C 1961 *Publ. Astron. Soc. Japan* **13** 450

# Chapter 17

## Star formation

## 17.1 Introduction

The arms of spiral galaxies are delineated by bright blue massive stars. Since these stars have short lifetimes, this indicates that star formation is an ongoing process in spiral galaxies. The spiral arms are also where large dark clouds of gas and dust, called *giant molecular clouds* (GMCs), are found, suggesting that the spiral density wave is responsible for compressing the interstellar medium (ISM) and triggering star formation. The ISM contains a number of phases which differ in density, temperature, and volume filling factor. Typical parameters are given in table 17.1, where $n$ is the atomic number density.

Note that the various ISM phases are in approximate pressure balance, i.e. $nT$ is a constant.

It is currently believed that star formation in our Galaxy occurs exclusively in the GMCs, as in the Orion molecular cloud complex. GMCs have typical masses of $10^4$–$10^6$ $M_\odot$ and can be hundreds of light years across. The average particle densities are $10^2$–$10^3$ particles cm$^{-3}$ but in the dense cores the density is $10^4$–$10^6$ particles cm$^{-3}$.

Because the GMCs are cold they emit very little radiation in the visible part of the spectrum but can be quite spectacular in the infra-red.

## 17.2 The Jeans mass

Star formation is a complex, multi-dimensional process, which is actively studied by computer modeling (see e.g. [1, 2]). It is not possible to describe here in any detail all of the physical processes involved (some books on the subject are [3–5]). Instead we consider a naïve analytical approach involving the Jeans mass, which is a measure of the size of the smallest structures in a cloud of dust and gas that will contract due to their self-gravity.

Before trying to estimate the Jeans mass, it is instructive to consider some properties of sound waves in the absence of gravity. Suppose we have a uniform

**Table 17.1.** Phases of the ISM.

| ISM phase | $n$ (atoms cm$^{-3}$) | $T$ (K) | Filling factor |
|---|---|---|---|
| GMCs | $10^2$–$10^6$ | 10–100 | 0.01 |
| Warm neutral medium | 0.2–0.5 | $6 \times 10^3$–$10^4$ | 0.1–0.2 |
| Warm ionized medium | 0.2–0.5 | $8 \times 10^3$ | 0.2–0.5 |
| Hot ionized medium | $10^{-4}$–$10^{-2}$ | $10^6$–$10^7$ | 0.3–0.7 |

medium of temperature $T$ and density $\rho$. If a sound wave travels through the medium, at any instant there will be regions where the density is increased (compression) and regions where the density is decreased (rarefaction) compared to the initial density. The work carried out by the pressure force will heat the compressed regions and cool the rarefied regions. Hence there will be heat flow from the compressed regions to the rarefied regions. If the thermal diffusivity is large enough, this heat flow will keep the temperature of the medium uniform at its initial value. This will be the case for the low density clouds, and so we will assume that the temperature does not change. The evolution of properties of the medium can be found from the continuity and momentum equations. Let the deviation of the density from its initial values be $\delta\rho(x, t)$. Provided the amplitude of the sound wave is small, the continuity equation gives

$$\frac{\partial}{\partial t}\delta\rho + \rho\nabla \cdot \mathbf{v} = 0, \tag{17.1}$$

where $\mathbf{v}(x, t)$ is the velocity of the material.

Assuming an ideal gas equation of state, the pressure deviation is

$$\delta p = \frac{\Re T}{\mu}\delta\rho = c_s^2\delta\rho, \tag{17.2}$$

where, as we will see later, $c_s$ is the phase velocity of the sound wave.

The conservation of momentum equation then gives

$$\rho\frac{\partial \mathbf{v}}{\partial t} = -\nabla\delta p = -c_s^2\nabla\delta\rho. \tag{17.3}$$

Eliminating the velocity from equations (17.2) and (16.4) gives a wave equation for the density perturbation

$$\frac{\partial^2\delta\rho}{\partial t^2} = c_s^2\nabla^2\delta\rho. \tag{17.4}$$

To make the analysis simpler, consider a plane wave traveling in the $x$-direction, so that $\delta\rho$ is a function of $t$ and a single Cartesian coordinate $x$. The wave equation is now

$$\frac{\partial^2\delta\rho}{\partial t^2} = c_s^2\frac{\partial^2\delta\rho}{\partial x^2}. \tag{17.5}$$

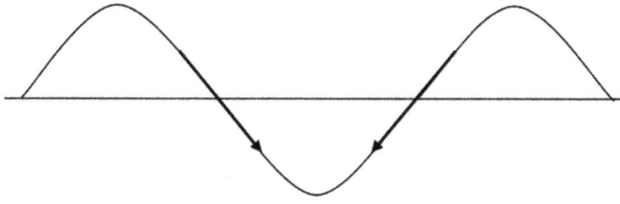

**Figure 17.1.** Schematic of the density profile in a standing wave.

The solution for $\delta\rho$ is a linear superposition of

$$\frac{\sin}{\cos}\left(kx \pm \omega t\right), \tag{17.6}$$

where

$$\omega = kc_s. \tag{17.7}$$

These solutions describe harmonic traveling waves of period $2\pi/\omega$, wavelength $2\pi/k$ and phase velocity

$$v_p = \frac{\omega}{k} = c_s. \tag{17.8}$$

Hence $c_s$ is the *isothermal sound speed*.

Since the molecular clouds have temperatures of order 10–100 K, typical isothermal sound speeds are 200–600 m s$^{-1}$.

Figure 17.1 shows schematically a snapshot of the density profile in a (standing) harmonic wave. The horizontal line represents the density of the unperturbed medium. The arrows show the direction of the pressure force at positions where the density deviation is zero. Hence, material moves away from high density regions towards the low density regions. In the absence of dissipation, the solution will overshoot the equilibrium state, leading to the oscillation.

Now let us add gravity to the picture. The direction of the gravitational force will be from the low density region to the high density region, i.e. in the opposite direction to the pressure force. If the gravitational force is larger than the pressure force, material will move from low density regions to the high density regions and will cause clumping of the material.

With gravity added, the momentum equation is

$$\rho\frac{\partial}{\partial t}\mathbf{v} = -c_s{}^2\nabla\delta\rho + \rho\mathbf{g}. \tag{17.9}$$

Here $\mathbf{g}$ is the gravitational acceleration arising from the density deviation. In our one-dimensional model

$$\frac{\mathrm{d}g}{\mathrm{d}x} = 4\pi G\delta\rho, \tag{17.10}$$

where, to be consistent with former use, we take $\mathbf{g} = -g\hat{\mathbf{x}}$, (but note that $g$ can be positive or negative).

For definiteness, take the spatial dependence of $\delta\rho$ to be

$$\delta\rho = \cos kx. \tag{17.11}$$

Then

$$g = \frac{4\pi G}{k} \sin kx. \tag{17.12}$$

The pressure force (per unit volume) is

$$-\frac{d\delta p}{dx} = -c_s^2 \frac{d\delta\rho}{dx} = c_s^2 k \sin kx. \tag{17.13}$$

The net force density is

$$-\frac{d\delta p}{dx} - \rho g = c_s^2 k \sin kx - \rho \frac{4\pi G}{k} \sin kx = \left( c_s^2 k - \frac{4\pi G\rho}{k} \right) \sin kx. \tag{17.14}$$

Hence the gravitational force is stronger than the force from the pressure gradient if

$$c_s^2 k^2 = \omega^2 < 4\pi G\rho. \tag{17.15}$$

Since $k$ is inversely proportional to wavelength, we see that gravity is unimportant for short wavelength, high frequency perturbations, but dominates for perturbations of wavelength greater than

$$\lambda > \lambda_J = \sqrt{\frac{\pi c_s^2}{G\rho}}. \tag{17.16}$$

From equation (17.15), we see that another way to phrase this condition is that gravity is unimportant if the period of the wave is much less than the dynamical time scale. A physical interpretation is that pressure cannot prevent gravitational contraction of a structure if the contraction time scale is shorter than the time for the pressure wave to communicate from one edge of the structure to the other.

The relevance of this analysis to star formation is that small cloud structures will not undergo gravitational contraction and hence cannot form stars. Equation (17.16) gives the limiting length scale. The corresponding mass scale (for a spherically symmetric geometry) is called the *Jeans mass*, given by

$$M_J = \frac{4\pi}{3}\rho\lambda_J^3 = \frac{4\pi}{3}\rho\left(\sqrt{\frac{\pi c_s^2}{G\rho}}\right)^3 = \frac{4\pi}{3}\left(\frac{\pi k}{G\mu m_u}\right)^{3/2}\frac{T^{3/2}}{\rho^{1/2}} \approx 4 \times 10^5 \left(\frac{T}{100\text{K}}\right)^{3/2} n^{-1/2} M_\odot, \tag{17.17}$$

where $n$ is the number density in atoms $\text{cm}^{-3}$.

The characteristic time scale for gravitational contraction is

$$\tau_J = \frac{\lambda_J}{c_s} = \sqrt{\frac{\pi}{G\rho}} \approx 10^8 \, n^{-1/2} \text{ years}. \tag{17.18}$$

**Table 17.2.** Jeans masses and contraction time scales for the various ISM phases.

| ISM phase | $n$ (atoms cm$^{-3}$) | $T$ (K) | Jeans mass ($M_\odot$) | Contraction time scale (years) |
|---|---|---|---|---|
| GMCs | $10^2$–$10^6$ | 10–100 | $10^1$–$10^5$ | $10^5$–$10^7$ |
| Warm neutral medium | 0.2–0.5 | $6 \times 10^3$–$10^4$ | $10^9$ | $10^8$ |
| Warm ionized medium | 0.2–0.5 | $8 \times 10^3$ | $10^9$ | $10^8$ |
| Hot ionized medium | $10^{-4}$–$10^{-2}$ | $10^6$–$10^7$ | $10^{14}$ | $10^9$–$10^{10}$ |

Hence densities greater than $10^{-4}$ atoms cm$^{-3}$ are required for the contraction time to be less than the age of the Universe.

Referring to table 17.1, we find Jeans masses of order 10 $M_\odot$ in the dense cores of molecular clouds. For the cloud as a whole, the Jeans mass is about $10^4$–$10^5$ $M_\odot$, which might be relevant to the formation of star clusters. Table 17.2 gives the Jeans mass and contraction time scale for the various ISM phases.

## 17.3 Fragmentation

Contraction will continue only as long as the energy released due to the decrease in gravitational potential energy can be radiated away. This will happen readily enough provided the cloud remains optically thin to infra-red radiation. Since the Jeans mass decreases as the density increases, smaller mass structures become Jeans unstable. Thus the cloud can fragment into smaller pieces provided the pieces themselves remain optically thin at infra-red wavelengths. Once the fragments become optically thick, radiation is trapped inside and the gas will heat up. This causes the Jeans mass to increase and fragmentation will end.

## Bibliography

[1] Bate M R 2009 *Mon. Not. R. Astron. Soc.* **392** 590
[2] Matsumoto T and Hanawa T 2011 *Astrophys. J.* **728** 47
[3] Stahler S W and Palla F 2005 *The Formation of Stars* (New York: Wiley)
[4] Hartmann L 2008 *Accretion Processes in Star Formation* (*Cambridge Astrophysics*) vol 47 (Cambridge: Cambridge University Press)
[5] Ward-Thompson D and Whitworth A P 2011 *An Introduction to Star Formation* (Cambridge: Cambridge University Press)

Structure and Evolution of Single Stars
An introduction
**James MacDonald**

# Chapter 18

# Evolution on the main sequence and beyond

## 18.1 Introduction

Stars on the MS evolve due to the changes in internal composition arising from the thermonuclear conversion of hydrogen into helium. Because this occurs on a time scale that is much longer than the star's thermal time scale, the star is in thermal equilibrium while it is on the MS.

## 18.2 Change in luminosity on the main sequence

The conversion of hydrogen into helium increases the mean molecular weight in the core of the star. If the density and temperature remained the same for a star in which radiation pressure is unimportant, the increase in molecular weight would lead to a decrease in the central pressure. To maintain hydrostatic balance, the central regions of the star adjust by contracting and heating. This leads to an increase in the total rate of energy generation and hence to an increase in the luminosity of the star.

Figure 18.1 shows the MS evolution in the HRD for solar composition stellar models of masses between 1 and 10 $M_\odot$. The broken line joins ZAMS models. We see that as the star evolves on the MS its luminosity increases. Also the two lower mass models evolve almost vertically in this diagram whereas the higher mass models evolve more horizontally towards lower effective temperatures. This difference in behavior is associated with the change from a radiative core at lower mass to a convective core at higher mass. Figure 18.2 is the same as figure 18.1 but for stellar models of masses between 10 and 100 $M_\odot$. We see that the relative luminosity increase is lower for the more massive stars. This is because radiation pressure, which is independent of molecular weight, aids in maintaining hydrostatic equilibrium.

doi:10.1088/978-1-6817-4105-5ch18     18-1

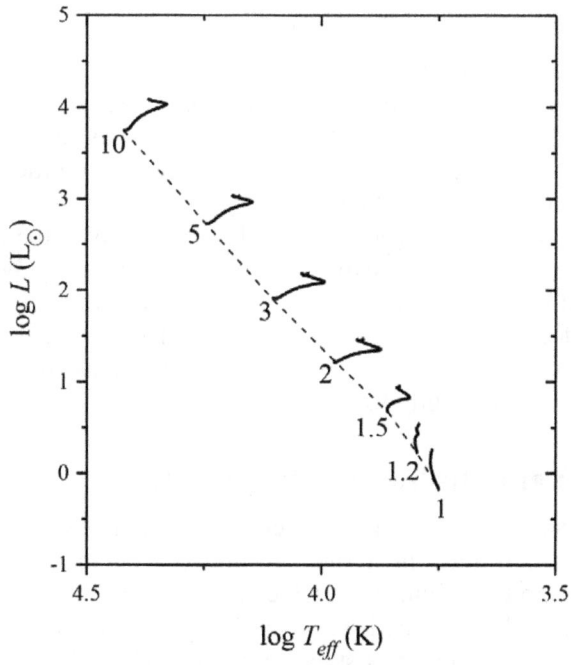

**Figure 18.1.** MS phase for stars of mass between 1 and 10 $M_\odot$.

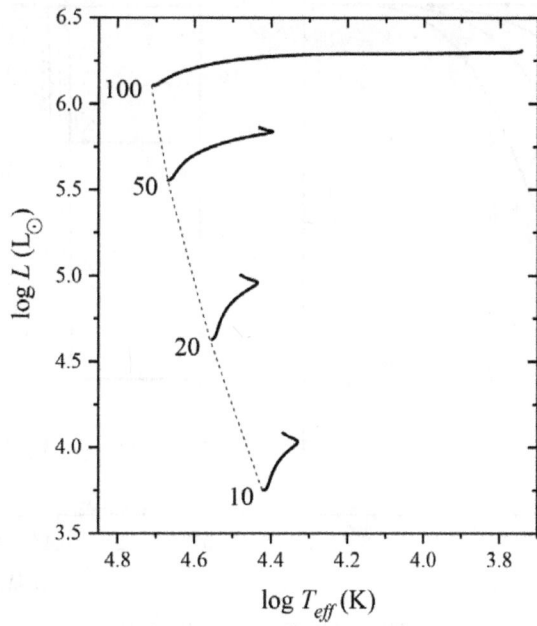

**Figure 18.2.** MS phase for stars of mass between 10 and 100 $M_\odot$.

## 18.3 Evolution of the hydrogen profile

Low mass and high mass stars differ in that the low mass stars have radiative cores whilst the high mass stars are convective in their cores. Because convection mixes the material, the hydrogen abundance remains uniform in the cores of massive stars while it is reduced by nuclear fusion. The difference in the evolution of the hydrogen profile is shown schematically in figure 18.3.

For the low mass star, hydrogen is depleted most rapidly at the very center of the star, whereas for the massive star hydrogen is depleted uniformly over the convective core. For the high mass star schematic, it has been assumed that the mass inside the convective core decreases with time. It is because of this structural difference that massive stars evolve to the red in the HRD whereas low mass stars evolve at almost constant temperature or slightly to the blue.

## 18.4 Evolution after hydrogen exhaustion in the core

In a low mass star ($M \leqslant 2.25 M_\odot$), after hydrogen is exhausted in the core, hydrogen burning continues in a shell around the helium core. The hydrogen burning adds mass to the helium core, which as a consequence contracts and heats. Initially, because it has no active nuclear energy source, the core is nearly isothermal and the contraction is slow. However, when the mass of the helium core reaches a critical value at which it cannot be supported against gravity by an

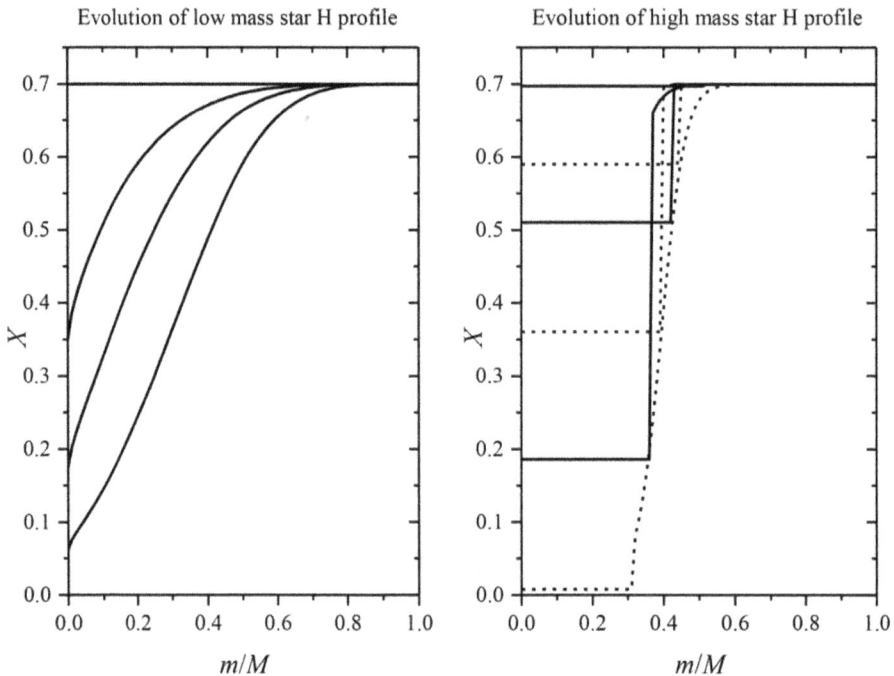

**Figure 18.3.** Schematic comparing the MS evolution of the hydrogen mass fraction profile in low mass and high mass stars.

isothermal ideal gas, the contraction speeds up [1]. This continues until, because of increasing density, the electrons in the core become degenerate. At this point the core contraction slows and the core is in a state in which degenerate electron pressure balances gravity.

The critical mass is called the Schönberg–Chandrasekhar limit. The existence of a critical mass can be demonstrated as follows. Let the mass, radius, and temperature of the isothermal core be $M_c$, $R_c$, and $T_c$, respectively. Multiplying the hydrostatic balance equation by $4\pi r^3$ and integrating over the core, we obtain

$$\int_0^{M_c} 4\pi r^3 \frac{dp}{dm} dm = -\int_0^{M_c} \frac{Gm}{r} dm. \tag{18.1}$$

Integrating the left-hand side by parts gives

$$\left[ 4\pi r^3 p \right]_0^{m=M_c} - 3\int_0^{M_c} \frac{p}{\rho} dm = -\int_0^{M_c} \frac{Gm}{r} dm. \tag{18.2}$$

Assuming an ideal gas, this gives for the pressure at the outer edge of the isothermal core

$$p_c = \frac{3}{4\pi} \frac{\Re T_c}{\mu_c} \frac{M_c}{R_c^3} - \frac{1}{4\pi R_c^3} \int_0^{M_c} \frac{Gm}{r} dm. \tag{18.3}$$

The integral on the right-hand side cannot be evaluated exactly without knowledge of the density distribution in the core. However, if the core evolves homologously, we can approximate the integral so that

$$p_c = \frac{3\Re}{4\pi\mu_c} \frac{M_c T_c}{R_c^3} - C_1 \frac{GM_c^2}{4\pi R_c^4}, \tag{18.4}$$

where $C_1$ is a constant of order unity.

For a given value of $M_c$ there is a maximum value of the pressure at the edge of the core, which can be found by differentiating $p_c$ with respect to $R_c$. The maximum pressure occurs at core radius

$$R_c = C_1 \frac{4}{9} \frac{\mu_c}{\Re T_c} GM_c, \tag{18.5}$$

and is

$$p_{c,max} = C_2 \frac{T_c^4}{M_c^2}, \tag{18.6}$$

where

$$C_2 = \left( \frac{9}{4C_1 G} \right)^3 \frac{3}{16\pi} \left( \frac{\Re}{\mu_c} \right)^4 \sim 10^{53} \text{ cgs.} \tag{18.7}$$

The pressure and temperature must be continuous at the boundary between the helium core and hydrogen-containing envelope. Making the further assumptions that the envelope evolves homologously and is also supported by gas pressure (but with a different molecular weight than the core), the pressure and temperature at the bottom of the envelope scale with stellar mass and radius as

$$p_e \propto \frac{M^2}{R^4}, \tag{18.8}$$

and

$$T_e \propto \frac{M}{R}. \tag{18.9}$$

Eliminating the stellar radius, and using continuity of pressure and temperature, this leads to

$$p_c = C_3 \frac{T_c^4}{M^2}, \tag{18.10}$$

where $C_3$ is a constant of order $10^{55}$ cgs.

Since $p_c$ must be less than $p_{c,\max}$, we must have

$$C_3 \frac{T_c^4}{M^2} \leqslant C_2 \frac{T_c^4}{M_c^2}. \tag{18.11}$$

As $M_c$ increases due to shell hydrogen burning, the right-hand side of the inequality decreases and the inequality will eventually be violated because $C_2 < C_3$. The core mass at which this occurs is the Schönberg–Chandrasekhar limit, $M_{SC}$. We see from equation (18.11), that $M_{SC} \propto M$.

The Schönberg–Chandrasekhar limit can be revealed by plotting how the central temperature evolves with core mass for a stellar model. This is shown in figure 18.4 for a 1 $M_\odot$ star. We see that $M_{SC} \approx 0.2 M_\odot$.

The scaling of the $M_{SC}$ with stellar mass has interesting implications for the higher mass stars. Since these stars have large convective cores, the mass of the helium core when it first forms will be higher than the $M_{SC}$.

Figure 18.5 shows for a 1 $M_\odot$ star how the central value of the ratio of pressure given by the non-relativistic degenerate electron formula to that from the non-degenerate electron formula evolves with the core mass. We see that the transition from non-degenerate to degenerate does occur when $M_c \approx M_{SC}$.

## 18.5 The Hertzsprung gap

In a low mass star, when the core mass first exceeds the Schönberg–Chandrasekhar limit, there is a relatively rapid decrease in effective temperature ($\Delta \log T_{\rm eff} \approx -0.05$) accompanied by increases in luminosity ($\Delta \log L \approx 0.2$) and radius ($\Delta \log R \approx 0.2$). These changes are shown for a 1 $M_\odot$ star in figures 18.6–18.8.

Hence the evolution in the HRD is up and to the red as shown in the figure 18.9. The filled circle is the approximate evolutionary point at which the core mass equals the Schönberg–Chandrasekhar mass.

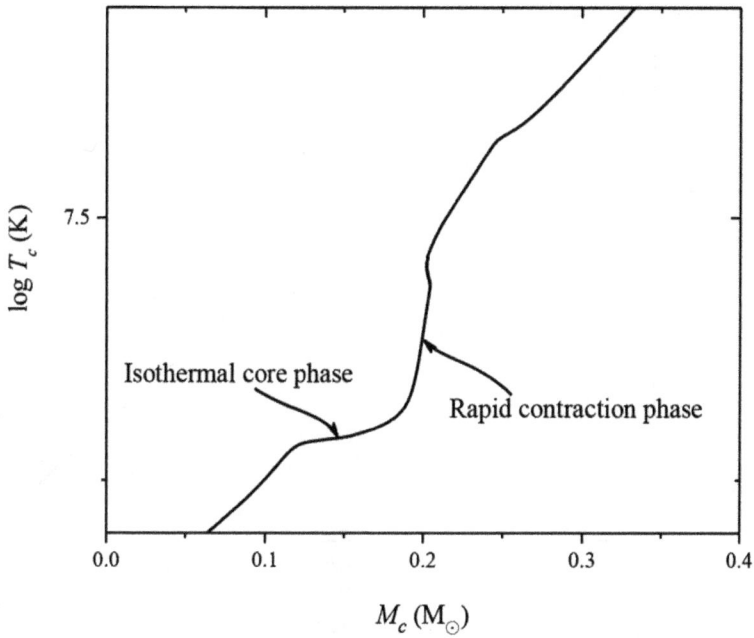

**Figure 18.4.** Evolution of the central temperature of a 1 $M_\odot$ model from the end of the MS to the beginning of the RGB. Here $M_c$ is the mass of the hydrogen exhausted core.

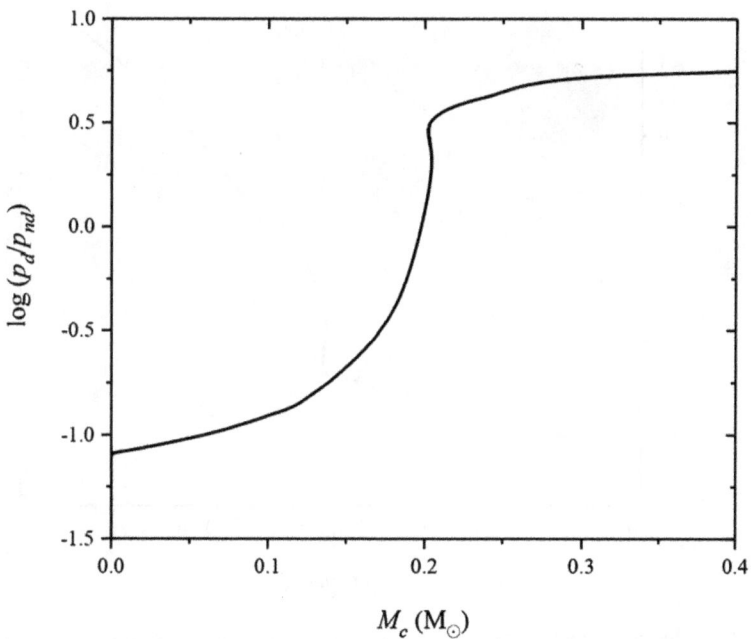

**Figure 18.5.** Evolution of the degenerate to non-degenerate electron pressure ratio at the center of a 1 $M_\odot$ model from the end of the MS to the beginning of the RGB.

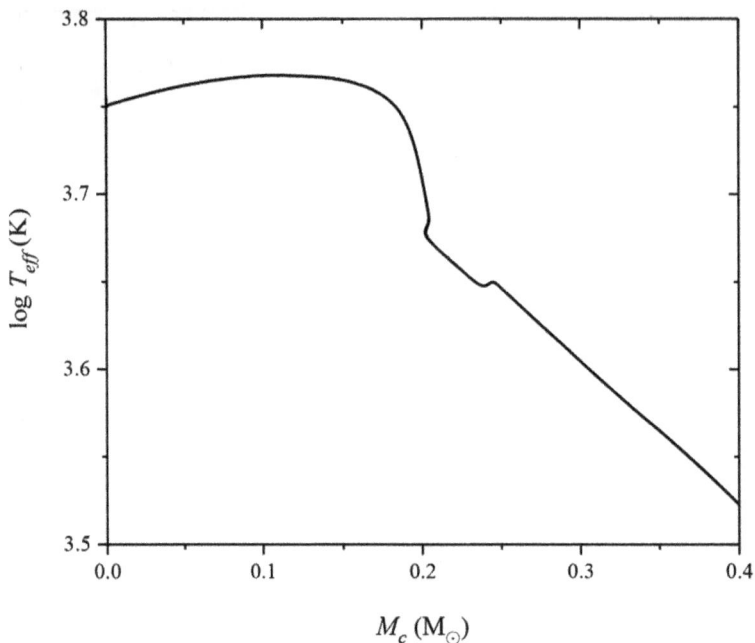

**Figure 18.6.** Evolution of the effective temperature of a 1 $M_\odot$ model from the end of the MS to the beginning of the RGB.

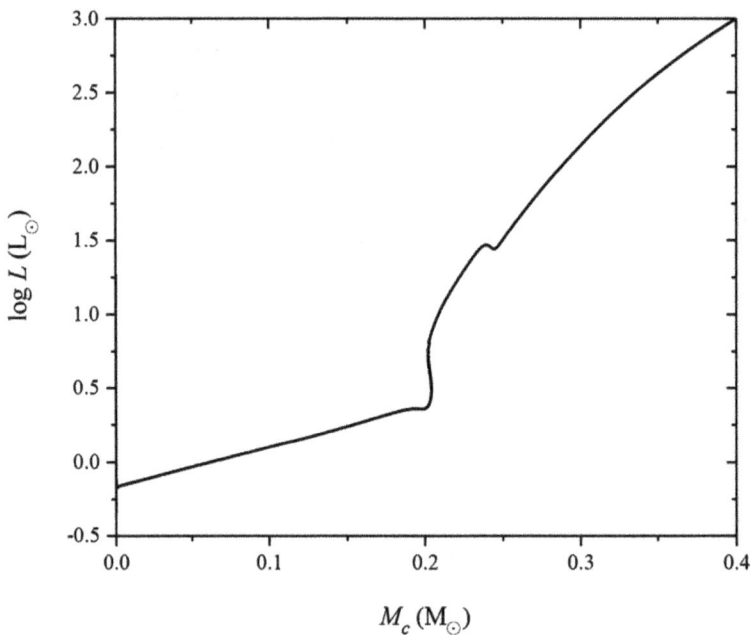

**Figure 18.7.** Evolution of the luminosity of a 1 $M_\odot$ model from the end of the MS to the beginning of the RGB.

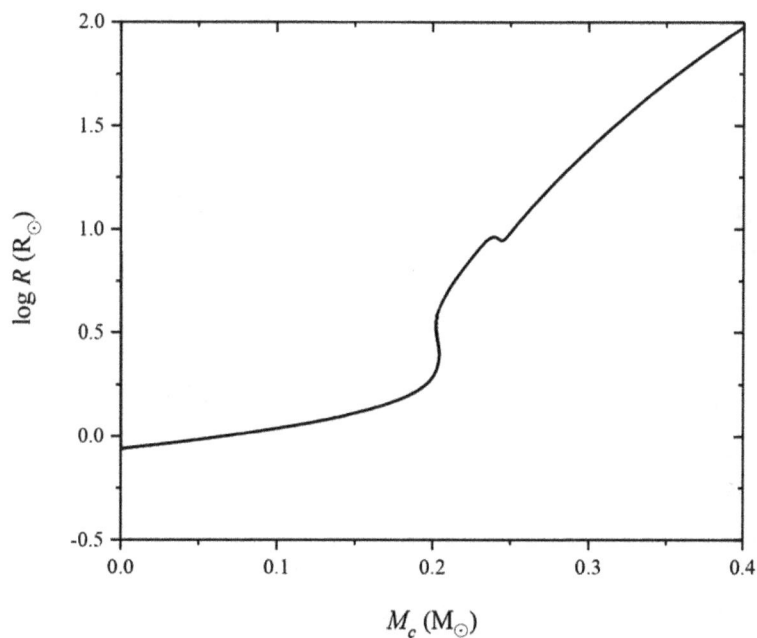

**Figure 18.8.** Evolution of the radius of a 1 $M_\odot$ model from the end of the MS to the beginning of the RGB.

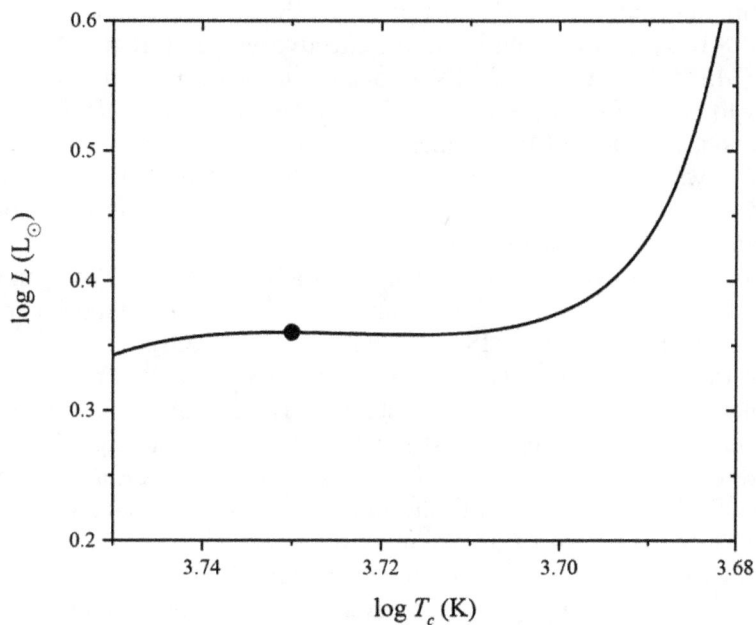

**Figure 18.9.** Evolution in the HRD of a 1 $M_\odot$ model from the end of the MS to the beginning of the RGB.

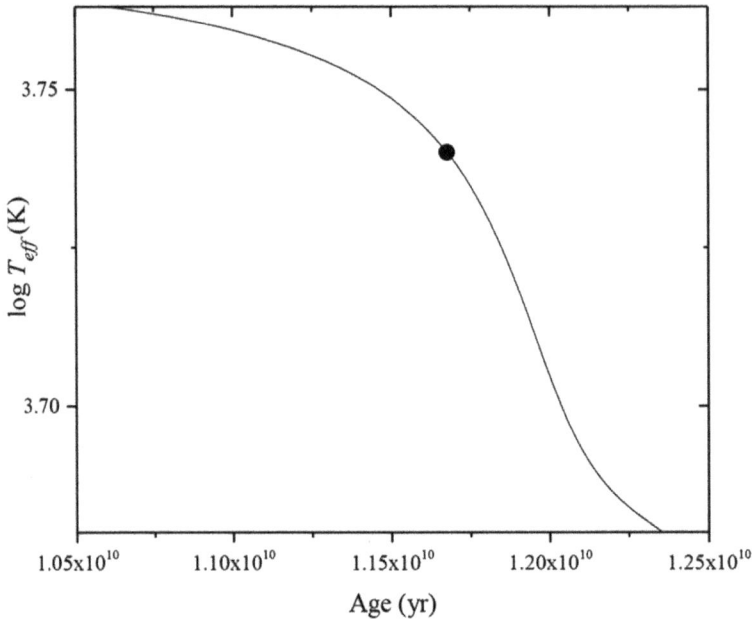

**Figure 18.10.** Change in the effective temperature of a 1 $M_\odot$ model with time from the end of the MS to the beginning of the RGB.

This portion of the evolutionary track begins at the end of the MS (i.e. at core H exhaustion) and continues to the start of the RGB.

Figure 18.10 shows the evolution of the effective temperature with time from the end of the MS to the bottom of the RGB for a model of a 1 $M_\odot$ star. The time for the 1 $M_\odot$ star to evolve from the end of the MS to the RGB is about 1.5 billion years, which is about 15% of the MS lifetime. For the most rapid part of this phase of the evolution, in which the star evolves across the HRD to the RGB, the time taken is about 0.5 billion years. Hence in a CMD of an old cluster, this portion of the isochrone will be well populated.

Now consider the evolution of a massive star after the end of hydrogen burning. Figure 18.11 shows how the effective temperature of a 10 $M_\odot$ model star changes with time from the end of the MS to the bottom of the RGB.

We see that the transition is rapid and occurs over a period of about 150 000 years, which is less than 1% of the MS lifetime. This short time scale is a result of the He core on formation having a mass greater than the Schönberg–Chandrasekhar limit. The core cannot be supported by isothermal gas pressure and contracts relatively quickly. The impact of this short time scale on a CMD of a young cluster is that this part of the isochrone will be sparsely populated relative to the MS.

Referring back to the CMD of the Hipparcos stars, figure 1.2 shown again here as figure 18.12, we see that a gap between the MS and RGB opens up with increasing luminosity. This gap is called the *Hertzsprung gap* and is also clearly seen for the

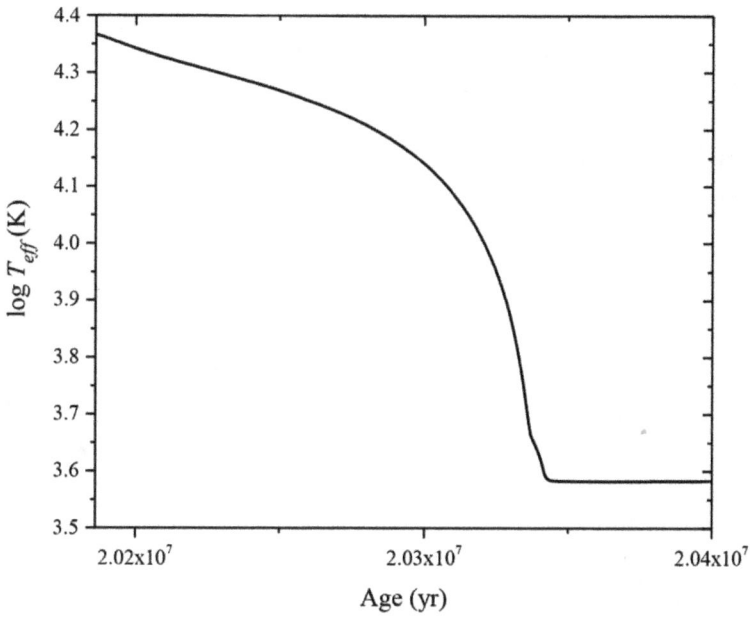

**Figure 18.11.** Change in the effective temperature of a 1 $M_\odot$ model with time from the end of the MS to the beginning of the RGB.

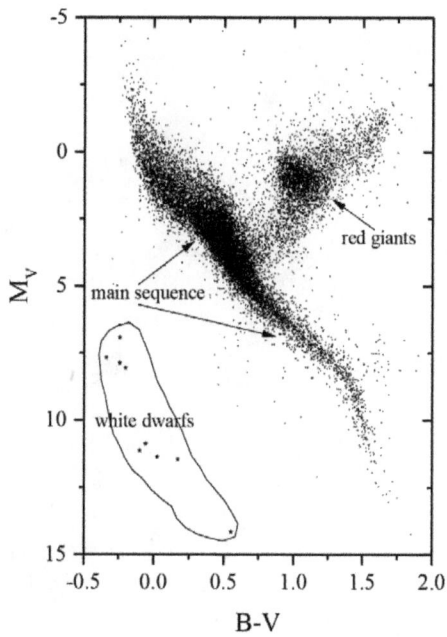

**Figure 18.12.** CMD for stars in the Hipparcos sample that have accurate parallaxes.

Hyades cluster in the composite cluster CMD (figure 1.4). This gap is a consequence of the difference in the MS–RGB transition time scale between stars that are radiative in the core on the MS and those that are convective.

## Bibliography

[1] Schönberg M and Chandrasekhar S 1942 *Astrophys. J.* **96** 161

Structure and Evolution of Single Stars
An introduction
**James MacDonald**

# Chapter 19

## Evolution on the red giant branch

### 19.1 Introduction

Stars on the RGB have helium cores which are not hot enough for helium burning to occur. The star is powered by hydrogen burning in a shell around the helium core. With the possible exception of some Pop III stars, it is the CNO-cycles that fuse hydrogen into helium. As the core grows in mass it contracts and heats. This increases the temperature of the burning shell and hence the hydrogen burning rate increases as the star evolves on the RGB. Because of their similarities to PMS Hayashi phase stars, RGB stars evolve at almost constant temperature and hence in CMDs the giant branch is almost vertical (see section 16.2).

### 19.2 Change in luminosity on the red giant branch

For stars of mass less than about 2.25 $M_\odot$, the electrons in the central regions of the core are degenerate, and it is degeneracy pressure that supports the inner core against gravity. The electrons in the outer core are non-degenerate and hence the core structure differs in detail from that of a WD. Nevertheless, it is a reasonable approximation to use the low mass WD mass–radius relation for the RGB star core. From equation (9.19), we have

$$M_c = 0.7 \left( \frac{R_c}{10^9 \text{ cm}} \right)^{-3} M_\odot. \qquad (19.1)$$

For $M_c > 0.1 \, M_\odot$, we find $R_c < \approx 2 \times 10^9$ cm, which is much smaller than the radius of the Sun, and much smaller still when compared to the radius of an RGB star. Hence we expect conditions at the edge of the core to be relatively insensitive to conditions at the surface of the star. In particular, we expect that the power of the hydrogen shell will depend on $M_c$ and $R_c$ more strongly than on the mass and radius of the star.

doi:10.1088/978-1-6817-4105-5ch19

Near the edge of the core, we assume that the density in the envelope can be approximated by

$$\rho = \rho_b \left( \frac{R_c}{r} \right)^\nu, \tag{19.2}$$

where $\rho_b$ and $\nu$ are constants. Assuming that the hydrogen burning shell is thin (in mass and radius), we can make the approximation that $m = M_c$ near the edge of the core. The hydrostatic balance equation then gives in the burning shell

$$p = \frac{GM_c \rho_b}{\nu + 1} \frac{R_c^\nu}{r^{\nu+1}}. \tag{19.3}$$

Since the electrons in the envelope are non-degenerate, the ideal gas law provides the temperature profile near the core edge and in particular in the hydrogen burning shell

$$T = \frac{\mu p}{\Re \rho} = \frac{\mu G M_c}{\Re(\nu + 1)} \frac{1}{r}. \tag{19.4}$$

The dependence of the luminosity of the star on the mass and radius of the core can be found by integrating the nuclear energy generation rate

$$L = \int_{R_c}^{\infty} \varepsilon_0 \rho T^\eta 4\pi r^2 \rho \, dr = \frac{4\pi R_c^3 \varepsilon_0 \rho_b^2}{2\nu + \eta - 3} \left[ \frac{\mu G M_c}{\Re(\nu + 1)R_c} \right]^\eta. \tag{19.5}$$

We cannot find $\rho_b$ without solving for the envelope structure. Although RGB stars have deep convective envelopes, the regions immediately above the hydrogen burning shell are radiative. Hence

$$L = -\frac{16\pi a c r^2 T^3}{3\kappa\rho} \frac{dT}{dr} = \frac{16\pi a c}{3\kappa\rho_b R_c^3} \left[ \frac{\mu G M_c}{\Re(\nu + 1)} \right]^4 \left( \frac{r}{R_c} \right)^{\nu-3}. \tag{19.6}$$

The luminosity depends on the opacity. If we assume electron scattering opacity, and that the hydrogen burning shell is thin so that $r$ can be replaced by $R_c$, then

$$L \propto \frac{M_c^4}{\rho_b R_c^3}. \tag{19.7}$$

Similarly, if instead we assume a Kramers' law opacity, then

$$L \propto \frac{M_c^{15/2}}{\rho_b^2 R_c^{13/2}}. \tag{19.8}$$

Eliminating $\rho_b$ from equations (19.5) and (19.7), we find that for electron scattering opacity

$$L \propto \frac{M_c^{(\eta+8)/3}}{R_c^{(\eta+3)/3}}, \tag{19.9}$$

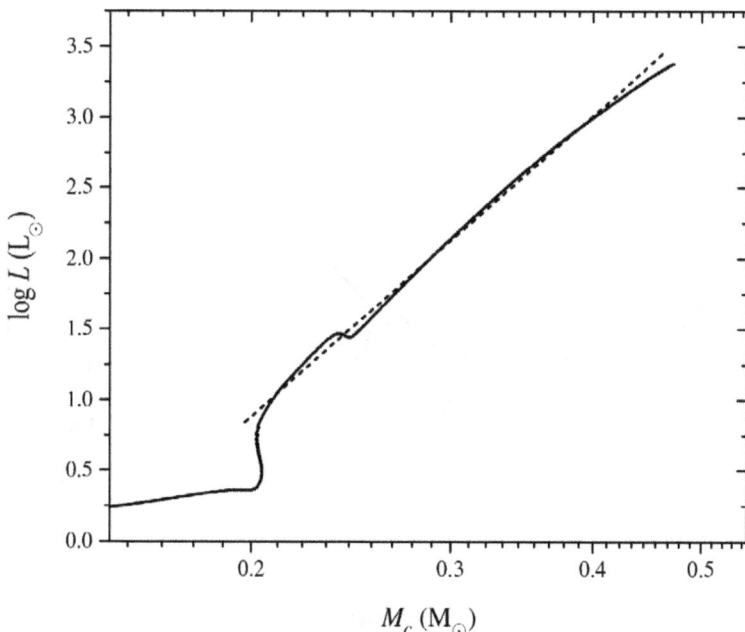

**Figure 19.1.** Evolution of the stellar luminosity with core mass up to the RGB tip for a 1 $M_\odot$ model.

which on using equation (19.1) gives

$$L \propto M_c^{4\eta/9+3}. \tag{19.10}$$

For Kramers' law of opacity, we find

$$L \propto \frac{M_c^{(2\eta+15)/4}}{R_c^{(2\eta+7)/4}} \propto M_c^{(2\eta+13)/3}. \tag{19.11}$$

Since for the CNO-cycles $\eta \approx 14$, we see that the luminosity is very sensitive to the mass of the core, with d log $L$/d log $M_c$ in the range 9–14.

Figure 19.1 shows the luminosity plotted against core mass for a 1 $M_\odot$ model. The dashed line has a slope of about 7, which indicates that the simple model overestimates the dependence of $L$ on the core mass. Even so, we do see a strong dependence. Figure 19.2 shows the luminosity plotted against core mass for models of mass 1, 1.5, and 2 $M_\odot$. We see that the luminosity on the RGB is insensitive to the mass of the star.

## 19.3 The globular cluster luminosity function bump

In figures 19.1 and 19.2, dips in luminosity can clearly be seen to occur as the star ascends the RGB. This dip also occurs as a kink in the HRD. Figure 19.3 shows the evolutionary track taken by a 1 $M_\odot$ model, starting at the ZAMS and up the lower part of the RGB. The inset is an enlargement of the region corresponding to the luminosity dip seen in figure 19.1.

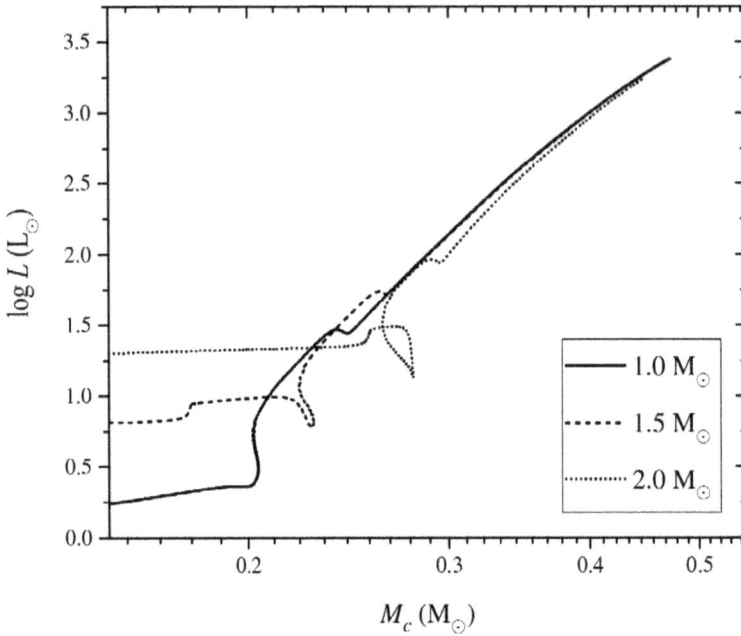

**Figure 19.2.** Evolution of the stellar luminosity with core mass up to the RGB tip for 1, 1.5, and 2 $M_\odot$ models.

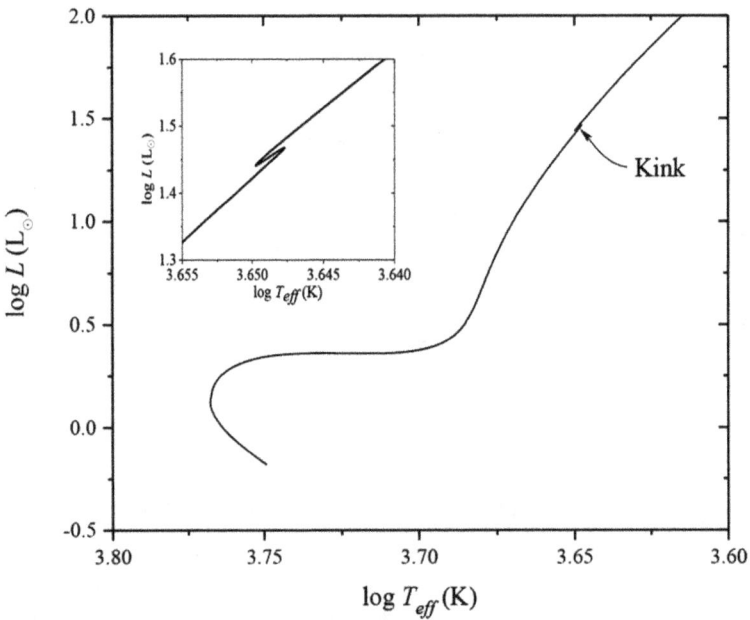

**Figure 19.3.** The RGB kink in the HRD for a 1 $M_\odot$ model.

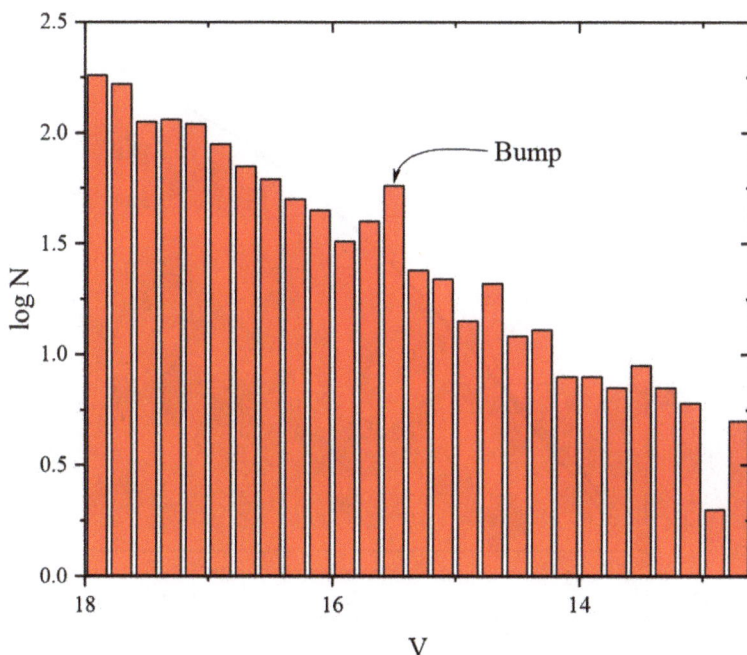

**Figure 19.4.** Luminosity function of the GC M3. Data from [1].

We see that during the kink the star, for a time, reverses its evolution. In a cluster with a large number of stars, such as a GC, the evolution through the kink shows up as a 'bump' in the luminosity function due to an excess of stars at the kink luminosity. This bump can be seen in figure 19.4 which shows the luminosity function for stars on the RGB of GC M3. The detection of observational evidence for the subtle bump feature is an important validation of stellar evolution theory.

The physical cause of the bump is the passage of the burning shell through a composition discontinuity left by the deepest extent of the surface convection zone. The shell moves into a region with a higher H mass fraction. This reduces the molecular weight and to maintain hydrostatic balance the temperature in the shell drops for a short time. The decrease in temperature causes a decrease in nuclear energy generation rate and consequently a decrease in the star's luminosity. Normal evolution resumes due to the increasing core mass.

## 19.4 The helium core flash

As a star evolves up the giant branch, its core grows in mass, which causes it to contract and heat. Eventually the core temperature becomes high enough for $3\alpha$ reactions to occur ($T \approx 10^8$ K). If the electrons are degenerate, then the pressure is insensitive to the temperature. The energy generated by the $3\alpha$ reactions heats the matter, further increasing the energy generation rate. This thermal instability is called the helium core flash and occurs in stars with an initial mass less than about $2.25\ M_{\odot}$. The increase in temperature ceases once the electrons become non-degenerate. For non-degenerate electrons at a fixed density, a temperature increase

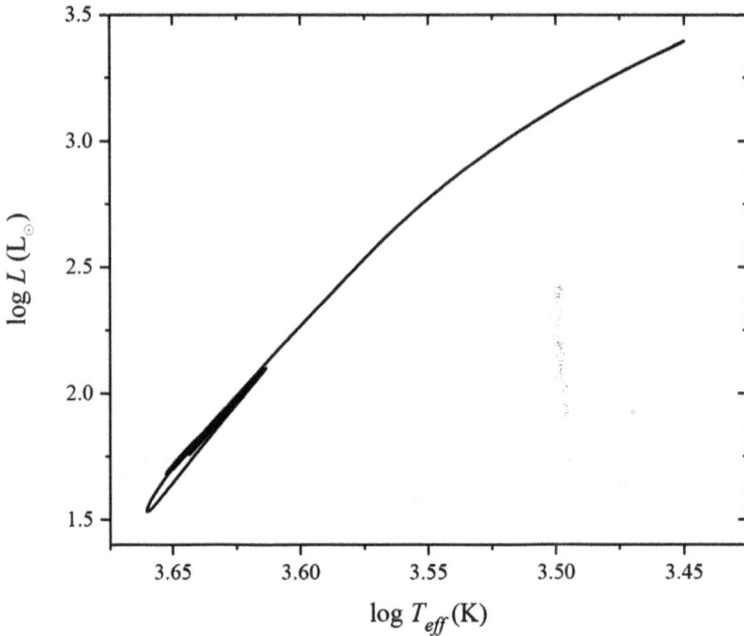

**Figure 19.5.** Evolution of a 1 $M_\odot$ model from the helium core flash to the beginning of central helium burning.

would cause an increase in pressure. Since a pressure increase causes expansion and cooling of the material, there is a feedback mechanism that leads to stability. At the peak of the core flash, the helium reactions produce energy at a rate in excess of $10^9\ L_\odot$. However, most of the energy initially goes into heating the core and, once degeneracy is lifted, into expanding the core.

Because of neutrino losses (see chapter 12), the core flash occurs off center. This causes an expansion and cooling of interior regions. Some of the heat from the flash diffuses inwards into the cooler, degenerate interior in a thermal front. This triggers a sequence of smaller flashes until the thermal front reaches the center.

The evolutionary track taken by a 1 $M_\odot$ model star from the start of the core flash to the point at which the thermal front reaches the stellar center is shown in figure 19.5.

The loops in the lower left are due to the smaller flashes after the main core flash. The short 1 Myr time scale for this phase of evolution can be seen from figure 19.6 in which the power of helium burning is plotted against the age of the star.

## 19.5 Stability considerations

To see why there is stability when the electrons are non-degenerate consider a simple uniform density model for the core. The core mass and radius are related by

$$M_c = \frac{4\pi\rho_c}{3}R_c^3. \tag{19.12}$$

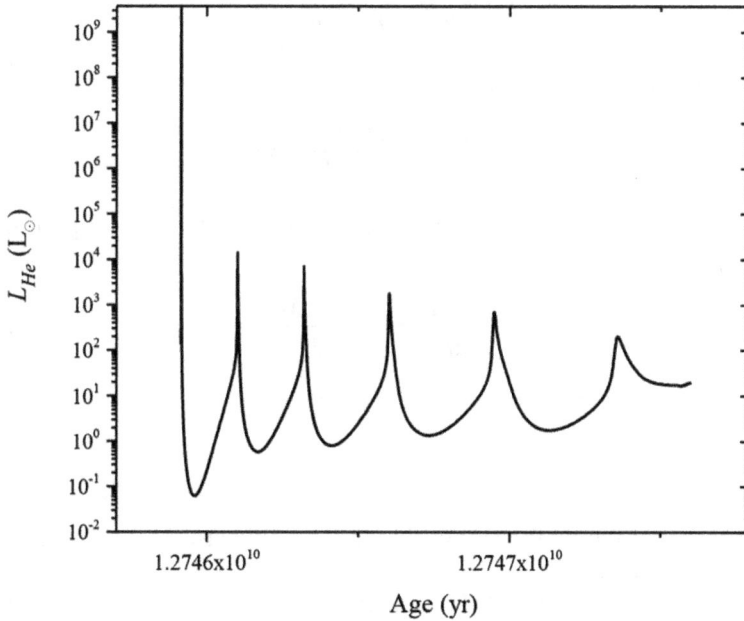

**Figure 19.6.** Luminosity from helium burning as a function of stellar age for the phase from the peak of the helium core flash to the start of central helium burning for a 1 $M_\odot$ model.

The central pressure is

$$P_c = \frac{2\pi G \rho_c^2}{3} R_c^2 = \frac{3G}{8\pi} \frac{M_c^2}{R_c^4}. \tag{19.13}$$

Ignoring energy transport (which also has a stabilizing effect), the energy equation is

$$\frac{dU}{dt} - \frac{p}{\rho^2} \frac{d\rho}{dt} = \varepsilon_{\text{nuc}}. \tag{19.14}$$

For an ideal gas equation of state, this becomes

$$\frac{3\Re}{2\mu} \frac{dT}{dt} - \frac{\Re}{\mu} \frac{T}{\rho} \frac{d\rho}{dt} = \varepsilon_{\text{nuc}}. \tag{19.15}$$

We also have from equation (19.13) that for an ideal gas

$$\frac{\Re T_c}{\mu} = \frac{2\pi G \rho_c}{3} R_c^2 = \frac{1}{2} \frac{GM_c}{R_c}. \tag{19.16}$$

Consider the two terms on the left-hand side of equation (19.15). If the core mass is constant, then from equations (19.12) and (19.16), we obtain that $\rho_c \propto T_c^3$. Hence the two terms have opposite sign and the second term is larger in magnitude. Since the right-hand side is positive, the left-hand side is also positive, which requires that the temperature decreases with time! This is a consequence of a star having negative specific heat if it is supported by ideal gas pressure.

Why does the temperature of a non-degenerate core increase with time? This is a consequence of the increasing core mass. Applying equation (19.15) at the center, and using equations (19.12) and (19.16), we obtain for the central temperature

$$-\frac{3\Re}{2\mu}\frac{dT_c}{dt} + \frac{2\Re T_c}{\mu M_c}\frac{dM_c}{dt} = \varepsilon_{\text{nuc}}. \tag{19.17}$$

Before helium ignition, the right-hand side is negligible and so (ignoring energy transport)

$$\frac{d\ln T_c}{dt} = \frac{4}{3}\frac{d\ln M_c}{dt}. \tag{19.18}$$

The temperature increases until

$$\frac{2\Re T_c}{\mu M_c}\frac{dM_c}{dt} \approx \varepsilon_{\text{nuc}}. \tag{19.19}$$

Note that a crucial factor for stability is that an increase in temperature leads to expansion and a decrease in pressure. Suppose instead that the pressure does not change. The energy equation (19.14) is then

$$C_p\frac{dT}{dt} = \varepsilon_{\text{nuc}}, \tag{19.20}$$

which shows that the temperature increases provided that the specific heat at constant pressure is positive, which it is for an ideal gas or a mixture of non-degenerate nuclei and degenerate electrons.

## Bibliography

[1] Rood R T *et al* 1999 *Astrophys. J.* **523** 752

# Chapter 20

## Evolution from red giant to white dwarf

### 20.1 Introduction

After the helium core flash, low mass stars experience a phase in which helium is converted to mainly carbon and oxygen in the central regions, and hydrogen is converted to helium in a shell around the helium core. This evolutionary phase corresponds to the HB part of the HRD. At the end of the HB, the radius and luminosity of the star both increase and the star evolves back to the giant branch. This evolutionary phase is called the asymptotic giant branch (AGB). The star leaves the AGB when the mass of the hydrogen-rich envelope has been decreased to about $10^{-3} M_\odot$ by a combination of hydrogen burning and wind mass loss. The star then evolves at roughly constant luminosity and increasing temperature. When the effective temperature reaches about 30 000 K the flux of ultra-violet photons becomes large enough to excite circumstellar material produced by mass loss, which might be seen as a planetary nebula. The effective temperature continues to increase until the envelope mass becomes too small to sustain nuclear reactions. During this phase the star is a central star of a planetary nebula (CSPN), at least until the planetary nebula disperses. After the cessation of nuclear reactions, the star cools and becomes a WD. Figure 20.1 shows the complete evolutionary path in the HRD from the PMS Hayashi phase to a cool WD. The star has initial mass 1 $M_\odot$ and heavy-element abundance $Z = 0.017$.

### 20.2 The horizontal branch

Figure 20.2 shows the RGB and HB parts of the evolutionary tracks taken in the HRD by 1 $M_\odot$ stars of heavy-element abundances $Z = 0.017$ and 0.00017. The transition phase from RGB to HB has been removed for clarity. Note that the HB of the $Z = 0.00017$ model is more clearly separated from the RGB than for the $Z = 0.017$ model, and hence the HB is easily seen in the CMDs of GCs (see figure 1.3).

We see that the HB stars are less luminous than the tip of the RGB. On the RGB the radiated power is supplied by the hydrogen burning shell. The helium core flash

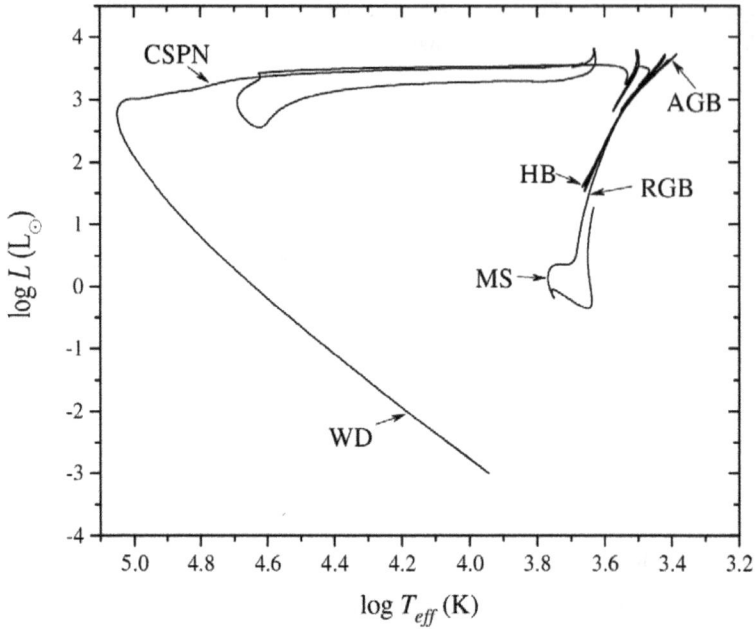

**Figure 20.1.** Complete evolutionary path in the HRD of a 1 $M_\odot$ model from the PMS to a cool WD.

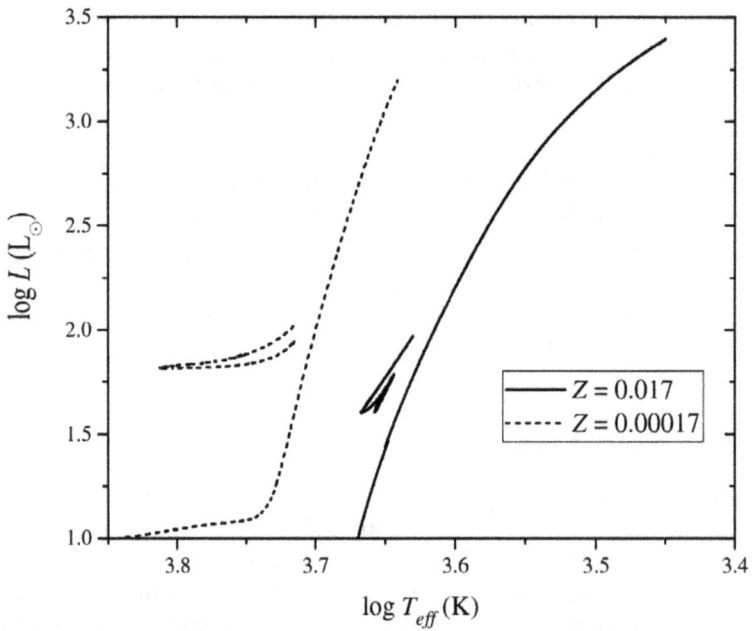

**Figure 20.2.** RGB and HB parts of the evolutionary tracks taken in the HRD by 1 $M_\odot$ models with $Z = 0.017$ and 0.00017.

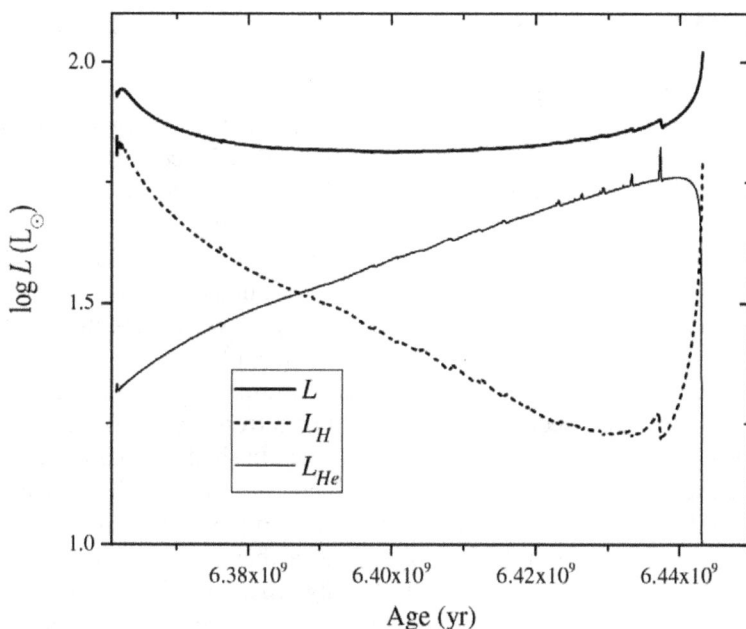

**Figure 20.3.** Luminosity evolution of a model of initial mass 1 $M_\odot$ and heavy-element abundance $Z = 0.00017$ when it is on the HB, shown together with the contributions from H and He burning.

results in an expansion of the core and a reduction in the temperature of the hydrogen shell. This reduces the luminosity from hydrogen burning. The HB helium burning luminosity is comparable to that of a helium MS star of mass equal to the mass of the HB star's helium core. Because the properties of the red giant core are most strongly dependent on the core mass, the core mass at helium ignition is independent of the star's mass and is about 0.47 $M_\odot$ for $Z = 0.017$. A helium ZAMS star of this mass has a luminosity of about 10–20 $L_\odot$, which sets the lower limit to the HB luminosity. The luminosity of a HB star is typically 50–100 $L_\odot$, which in part comes from core helium burning and in part from shell hydrogen burning. Figure 20.3 shows the luminosity of a star of initial mass 1 $M_\odot$ and heavy-element abundance $Z = 0.00017$ when it is on the HB as a function of time, together with the contributions from H and He burning.

## 20.3 The asymptotic giant branch

Because of the strong dependence of the rates of helium burning reactions on temperature, the HB stars have convective cores. Helium exhaustion occurs over an extended region of the core and hence the helium exhausted core contracts relatively rapidly until the electrons become degenerate. The contraction causes an increase in the temperature of the helium burning shell and consequently an increase in the rate of helium burning. The star increases in luminosity and also increases in radius to red giant dimensions. Because the evolutionary track becomes tangential to the RGB, this phase is called the AGB. In the early stages of the AGB, the star has two shells where nuclear reactions occur. In the inner shell around the degenerate CO core, helium is converted mainly to C and O, which increases the core mass and causes it to contract. Further out is a shell in which H is converted to He. Initially the luminosity provided by nuclear burning in the He shell is larger than that of the H shell. Because more energy is produced per unit mass by converting H to He than by converting He to C or O, the mass of the He-rich layer between the two burning shells decreases with time and eventually the He burning luminosity is diminished. The He-rich layer then contracts so that the temperature in the H shell increases. The H burning luminosity then increases and the helium layer mass increases again. Added mass compresses and heats the helium layer causing re-ignition of the He burning shell. However because of the small scale height in the He shell compared to the radius of the core, the He shell is thermally unstable. This instability is called the thin shell instability and occurs even though the electrons are non-degenerate [1]. As a consequence of the near planar geometry in the He shell, to maintain hydrostatic balance the pressure in the He shell is determined only by the column density above the shell and does not change significantly as the temperature in the shell increases. For reasons given in section 20.5, this leads to a thermonuclear runaway called a *shell flash* or a *thermal pulse*, which ends only when the density in the shell has decreased sufficiently that the scale height in the shell becomes comparable to the radius of the core and curvature becomes important. Because of the expansion of the He layer, the H shell moves to larger radii and its temperature decreases to the point that H burning is extinguished.

The further evolution of the star consists of a sequence of thermal pulse cycles [2]. A complete cycle consists of a thermal pulse followed by a phase of quiescent helium burning which reduces the helium layer mass to the point that the He burning luminosity is diminished and the H shell is re-established. The H burning then increases the mass of the He layer. The added mass compresses and heats the layer so that cycle begins again. This evolutionary phase is called the thermally pulsing AGB (TPAGB). For a star of initial mass of 1 $M_{\odot}$, the early AGB phase lasts about $10^7$ years and a complete thermal pulse cycle takes about $10^5$ years. The number of cycles experienced by a star depends critically on the mass of the star and the rate at which mass is lost by stellar winds. The thermal pulses show up as loops in the HRD and can be seen in figure 20.1 between the parts labelled AGB and CSPN.

## 20.4 The formation of planetary nebulae

Planetary nebulae display a wide range of shapes and structures. Some are ring shaped whereas others are bipolar or even helical. The wide range in morphology suggests that there is more than one way to form or shape a planetary nebula.

One possible scenario [3] is that while on the AGB the star has a slow dense wind that is responsible for removing the bulk of the hydrogen-rich envelope. Empirical evidence for such winds is provided by observations of Mira variables and related stars [4–6]. Once the effective temperature becomes higher than $10^4$ K, it is likely that a radiation-driven wind similar to those of O stars occurs. Indeed, such winds have been detected in some CSPN [7–9]. Because of the increase in the depth of the gravitational potential well, this wind will have a higher terminal velocity than the wind from the cool AGB star. Hence it will catch up with the slow wind and compress it to form a shell. This shell becomes the planetary nebula when the central star becomes hot enough to produce ultra-violet photons to ionize it.

An alternative possibility is that the outer layers of the AGB star are ejected during the luminosity peak of a thermal pulse cycle due to a dynamical instability associated with hydrogen recombination that occurs when the luminosity exceeds a critical value [10]. In view of the large number of non-spherical planetary nebulae, a binary star formation channel is highly probable [11–13].

## 20.5 The cooling of white dwarfs

Before we consider WD cooling, let us reprise how WDs are formed. Figure 20.4 is the same as figure 20.1, except that now the evolutionary track is labelled with points at which various phases begin or end. Point 1 marks the start of the evolutionary calculation on the PMS phase. Points 2 and 3 mark the start and end of the core hydrogen burning MS phase. Point 4 marks the start of the core helium flash that ends the red giant phase. Point 5 marks the start of quiescent core helium burning. After the end of core helium burning (point 6), helium burns in a shell, stably at first, but once the region between the helium and hydrogen burning shells become thin, helium shell flashes (i.e. thermal pulses) occur. Three of the He shell flash loops can be seen in the diagram, labeled A, B, and C. Point 7 marks where the nuclear energy production ends and the WD cooling track begins. At the end of the evolution, mass loss has reduced the stellar mass to 0.541 $M_\odot$.

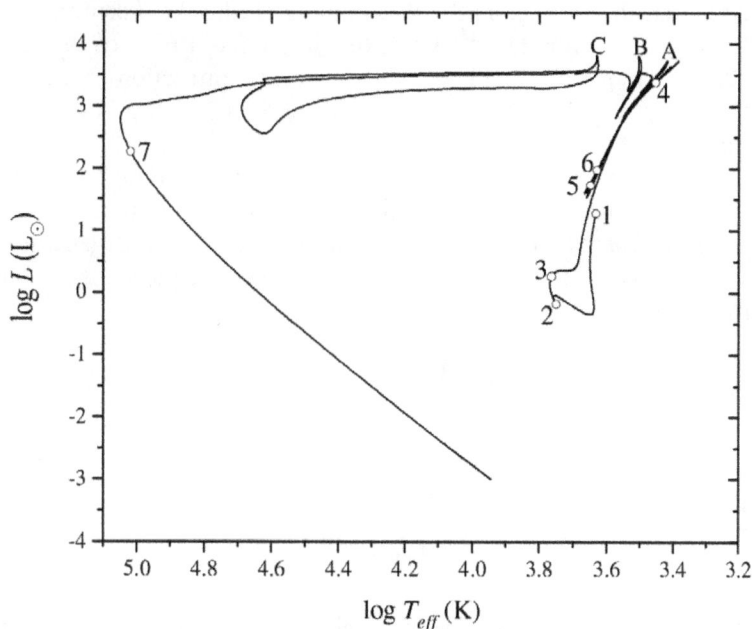

**Figure 20.4.** Complete evolutionary path in the HRD of a 1 $M_\odot$ model from the PMS to a cool WD.

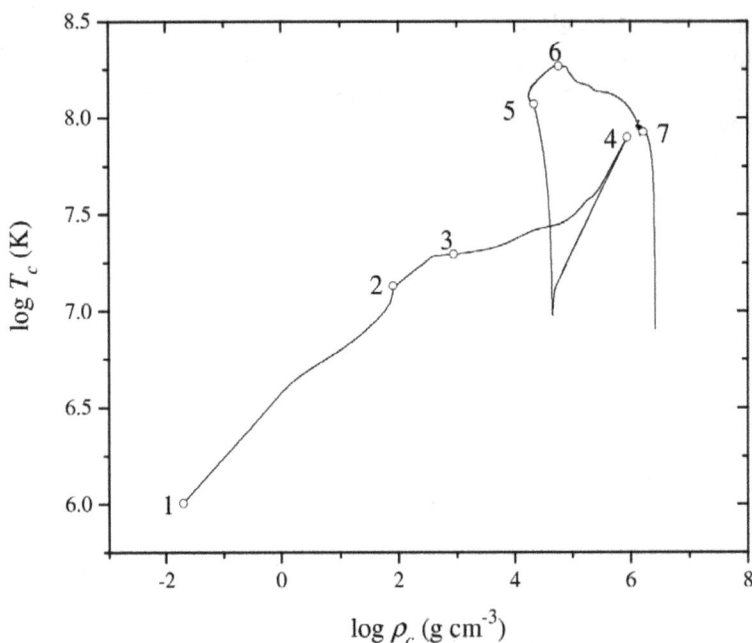

**Figure 20.5.** Complete evolutionary path in the central temperature–central density plane of a 1 $M_\odot$ model from the PMS to a cool WD.

Figure 20.5 shows how the central temperature and density change through the evolution of the star. The numbers mark the same evolutionary points as in the first figure. We see that the star enters the WD cooling track with a central temperature of $8.4 \times 10^7$ K. At this point, the central density is about 63% of the final central density of $2.62 \times 10^6$ g cm$^{-3}$ and hence there is some contraction of the core during cooling. The composition at the center is, by mass, 0.665 $^{16}$O, 0.313 $^{12}$C, and 0.022 heavier elements.

Are the electrons degenerate in the core at the start of cooling? If they are, then they are non-relativistic because the density is too low for relativistic electrons. The quick test for degeneracy is to compare the expression for degenerate electron pressure with that for non-degenerate electrons. These give the same pressure when, in cgs units,

$$10^{13}\left(\frac{\rho}{\mu_e}\right)^{5/3} = 8.3 \times 10^7 \frac{\rho}{\mu_e}T, \qquad (20.1)$$

which gives a transition temperature of

$$T_{\mathrm{nd\_d}} = 1.2 \times 10^5 \left(\frac{\rho}{\mu_e}\right)^{2/3}. \qquad (20.2)$$

The values for density and temperature given above give that the electrons are degenerate ($T < 0.1 T_{\mathrm{nd\_d}}$) and become more so as the core temperature decreases.

The first quantitative study of the rate of cooling of WDs was made by Mestel [14]. A WD has an electron degenerate core with a thin non-degenerate envelope ($m \approx 10^{-4} \, M_\odot$). In the core the degenerate electrons have a large mean free path because almost all available energy levels in the Fermi 'sea' are filled. This results in a high thermal conductivity and the core is isothermal to a high degree. We can assign a single temperature, $T_c$, to the core. Because the WD is supported by degenerate electron pressure very little energy can be released by gravitational contraction (except during the very early phases). Also very little energy can come from the thermal energy of the electrons because most of them are already in the lowest energy states. In addition, essentially all nuclear processes are finished (but hydrogen burning can linger on [15]). Hence the major source for the energy emitted from the surface in photons or neutrinos comes from cooling of ions in the core. (In the early stages of WD cooling, thermal neutrino losses are important, but because they scale like $T^{15/2}$ they rapidly become negligible.) The stellar luminosity is then related to the central temperature by

$$L_* = -M C_V \frac{dT_c}{dt},$$ (20.3)

where $C_V = 3\Re/2A$ is the specific heat per unit mass for a monatomic gas of atomic weight $A$.

To obtain a second relation between $L_*$ and $T_c$, the non-degenerate envelope is assumed to be radiative, with Kramers' opacity law

$$\kappa = \kappa_0 \rho T^{-7/2}.$$ (20.4)

The envelope is thin with very little energy generation, so that in the envelope $L = L_*$, and $m = M$ to a good approximation. From the equations of radiative transfer and hydrostatic equilibrium

$$\frac{dT}{dp} = \frac{3\kappa L_*}{16\pi ac T^3 GM}.$$ (20.5)

Using (20.4) and an ideal gas equation of state, this gives

$$\frac{dT}{dp} = \frac{3\kappa_0 \mu L_*}{16\pi ac \Re GM} p T^{-15/2},$$ (20.6)

which for zero boundary conditions at the surface gives

$$T^{17/2} = \frac{17}{4} \frac{3\kappa_0 \mu L_*}{16\pi ac \Re GM} p^2$$ (20.7)

(from which we can deduce that the envelope is radiative, provided there are no ionization zones).

The core–envelope interface is where electrons become degenerate (see section 8.3), i.e. where

$$K_1 \left( \frac{\rho}{\mu_e} \right)^{5/3} = \frac{\Re}{\mu_e} \rho T.$$ (20.8)

In terms of temperature and electron pressure this expression is, in cgs units,

$$p_e = 2.0 T^{5/2}. \tag{20.9}$$

The total pressure (ion plus electron) at the interface is

$$p = p_e + p_{ion} = \left(1 + \frac{\mu_e}{\mu_{ion}}\right) p_e = \frac{\mu_e}{\mu} p_e. \tag{20.10}$$

Hence

$$p = 2.0 \frac{\mu_e}{\mu} T^{5/2}, \tag{20.11}$$

at the core–envelope interface. Inserting this into equation (20.7) gives the central temperature. Putting in the numerical values for the constants, we find

$$L_* = 4 \times 10^7 T_c^{7/2} \frac{M}{M_\odot}. \tag{20.12}$$

Using this with equation (20.3), we obtain

$$\frac{dT_c}{dt} = -1.6 \times 10^{-34} T_c^{7/2}, \tag{20.13}$$

which has solution

$$T_c = 5.8 \times 10^6 \left(\frac{t}{10^9 \text{ years}}\right)^{-2/5}. \tag{20.14}$$

From equation (20.12), we find

$$\frac{L_*}{L_\odot} = 4.67 \times 10^{-3} \frac{M}{M_\odot} \left(\frac{t}{10^9 \text{ years}}\right)^{-7/5}. \tag{20.15}$$

In table 20.1, the cooling times for a 0.6 $M_\odot$ WD from Mestel theory are compared with the cooling times from detailed calculations [15].

## 20.6 The luminosity function of white dwarfs

The WD luminosity function from the Sloan Digital Sky Survey [16] is shown in figure 20.6. Here $N$ is defined such that the space density of WDs per unit interval of $M_{Bol}$ is $N(M_{Bol})dM_{Bol} pc^{-3} M_{Bol}^{-1}$. The integral of the luminosity function gives the

**Table 20.1.** WD cooling times.

| $L_*/L_\odot$ | Mestel age (years) | IM85 age (years) |
|---|---|---|
| $10^{-2}$ | $4 \times 10^8$ | $2.5 \times 10^8$ |
| $10^{-3}$ | $2 \times 10^9$ | $10^9$ |
| $10^{-4}$ | $10^{10}$ | $5 \times 10^9$ |

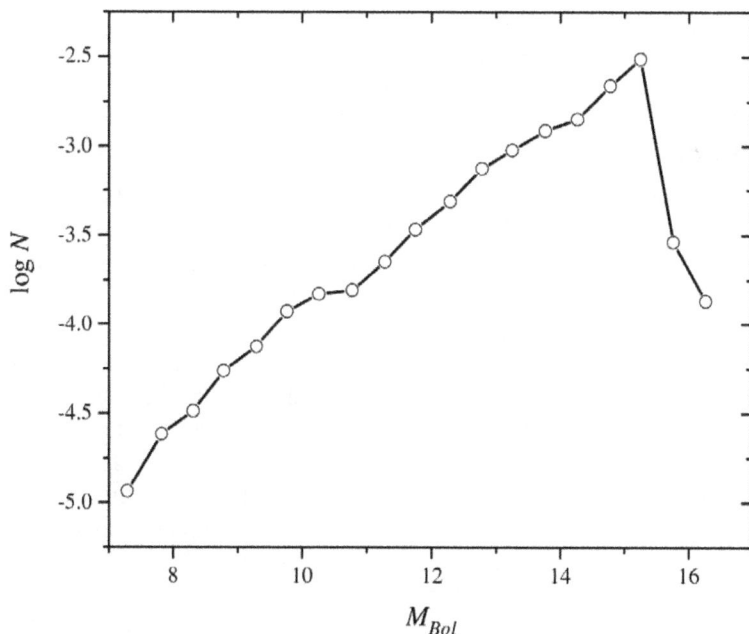

**Figure 20.6.** WD luminosity function from the Sloan Digital Sky Survey [16].

space density of WDs in the solar neighborhood, 0.005 pc$^{-3}$, which indicates that about 1 in 10 stars in the solar neighborhood is a WD.

The average slope of the luminosity function before the sharp drop at $M_{\text{Bol}} = 15.4$ is d log $N$/d$M_{\text{Bol}} = 0.291$. The bolometric magnitude is related to luminosity by

$$M_{\text{Bol}} = 4.755 - 2.5 \log\left(\frac{L}{L_\odot}\right). \tag{20.16}$$

Hence d log $N$/d log $L = -0.727$. If we assume that WDs form at a uniform rate and that they all have the same mass and composition so that they all follow the same cooling curve, then the number of WDs in any bin will be proportional to the lifetime in that bin. With these assumptions, the Mestel cooling law predicts d log $N$/d log $L = -5/7 = -0.714$, which is remarkably close to the observed value, considering the simplicity of the assumptions. Large deviations from the mean slope are often attributed to variations in the rate of formation of WD forming stars.

The sharp drop off in the luminosity function at $M_{\text{Bol}} = 15.4$ indicates a deficiency of WDs at luminosities below $5.5 \times 10^{-5} L_\odot$. If we use the simple Mestel cooling law, this corresponds to a cooling time of 16.5 Gyr, which is longer than the age of the Universe [17], 13.7 Gyr. If we divide by the factor 2 indicated by the IM85 calculations, the cooling time is reduced to 8.3 Gyr. If we accurately knew the masses of the coolest WDs and how these masses are related to the initial masses of the progenitor stars so that we can add the lifetime of the progenitor mass, we can determine a lower limit on the age of the Galaxy [18]. Complications arise from the chemical evolution of the Galaxy and the lack of a clear correlation between

Galactic age and chemical abundance, not to mention mixing of stars from different parts of the Galaxy and the effects of the essentially unknown stellar He mass fraction.

## 20.7 Masses of white dwarf stars: observational material

Accurate masses for WDs can be obtained from a few visual binary systems in which we can measure the orbit, e.g. Sirius A + B. The WD radius is then obtained from its luminosity and effective temperature. Data for WDs with accurate masses are given in table 20.2.

Less accurate masses can be obtained by spectroscopic methods. The widths of spectral lines and the number of distinguishable lines depend on the value of the surface gravity. Higher gravity leads to higher density in the photosphere, which means there are more frequent collisions between particles. Thus the wave trains of emitted photons tend to be shorter which leads to broader spectral lines. The strengths of spectral lines as well as the strength of continuum emission also depend on $T_{\rm eff}$. Hence analysis of the spectrum of a WD allows measurement of its surface gravity and effective temperature. The theoretical mass–radius relation, possibly with a correction for finite temperature, allows the WD mass to be obtained from its surface gravity. This method can be applied to a large number of single WDs, which permits statistical studies of WD masses. Histograms of surface gravity and derived mass are shown in figure 20.7. We see that the surface gravity distribution peaks near $\log g = 8$ and the mass distribution peaks near $M = 0.55\ M_{\odot}$. Note that there are very few WDs with mass $>1\ M_{\odot}$.

Measurement of the masses of WDs in open clusters with known ages allows determination of the relation between the mass of a WD and the mass of its MS progenitor. The WD cooling time is subtracted from the cluster age to find the time the star spends in the pre-WD stage, which allows the progenitor mass to be found from evolutionary models. Data for WDs in some open clusters and the GC M4 are shown in figure 20.8 together with some theoretical predictions [22]. We see that there is a large amount of scatter which in part is due to differences in the clusters' heavy-element abundances. The WDs with anomalously low mass were probably formed in interacting binary systems. For the other WDs, it is clear that some progenitor stars must have lost the majority of their mass before they became WDs. For progenitor masses $>3\ M_{\odot}$, the WDs are on average more massive than the

Table 20.2. Accurate WD masses. Data from [19, 20].

| Name | WD mass $(M_{\odot})$ | WD radius $(R_{\odot})$ | Companion mass $(M_{\odot})$ | Companion radius $(R_{\odot})$ |
|---|---|---|---|---|
| Sirius | 1.000 | 0.0084 | 2.12 | 1.711 |
| Procyon | 0.604 | 0.01234 | 1.497 | 1.7 |
| 40 Eri | 0.501 | 0.0136 | 0.89 | 0.85 |
| Stein 2051 | 0.66 | 0.011 | | |
| G107-70 | 0.65 | | | |

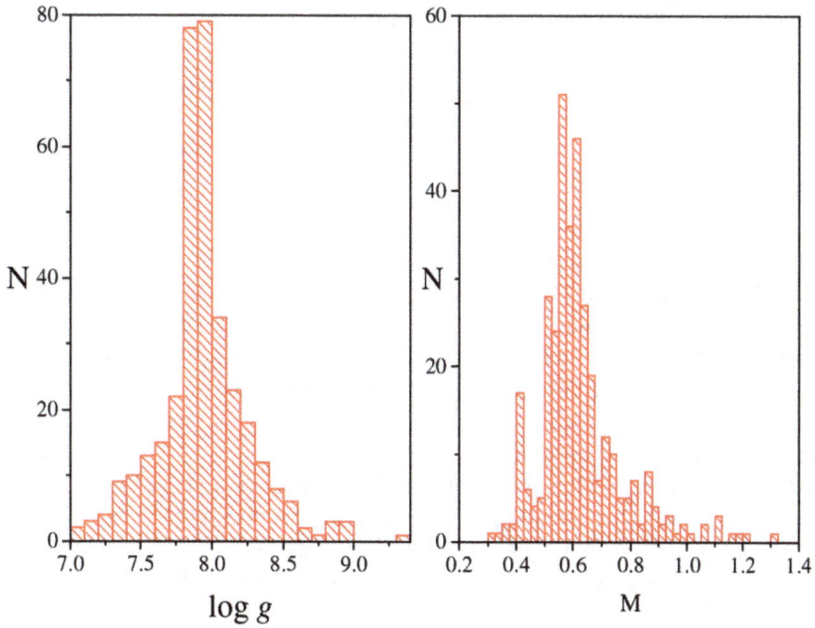

**Figure 20.7.** Surface gravity and mass distributions for 298 DA stars with $T_{\mathrm{eff}} > 13\,000$ K. Data from [21].

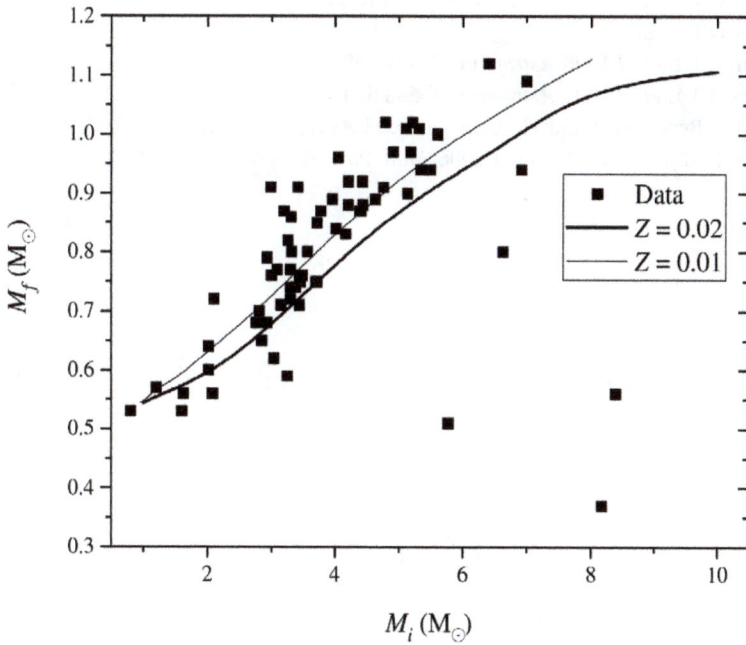

**Figure 20.8.** The WD initial mass–final mass ($M_i$–$M_f$) relation. Data points are from the literature. The lines are theoretical predictions from [22].

predicted values by about 0.1 $M_\odot$. This discrepancy is most probably due to a combination of the mass loss rates adopted for the modeling being too high and the neglect of convective core overshoot, which if included would give larger cores during hydrogen and helium core burning phases.

## Bibliography

[1] Schwarzschild M and Härm R 1965 *Astrophys. J.* **142** 855

[2] Iben I and Renzini A 1983 *Annu. Rev. Astron. Astrophys.* **21** 271

[3] Kwok S, Purton C R and Fitzgerald P M 1978 *Astrophys. J. Lett.* **219** L125

[4] Lepine J R D, Ortiz R and Epchtein N 1995 *Astron. Astrophys.* **299** 453

[5] Knapp G R, Young K, Lee E and Jorissen A 1998 *Astrophys. J. Suppl.* **117** 209

[6] Whitelock P A, Feast M W, van Loon J T and Zijlstra A 2003 *Mon. Not. R. Astron. Soc.* **342** 86

[7] Heap S R *et al* 1978 *Nature* **275** 385

[8] Modigliani A, Patriarchi P and Perinotto M 1993 *Astrophys. J.* **415** 258

[9] Guerrero M A and De Marco O 2013 *Astron. Astrophys.* **553** A126

[10] Wagenhuber J and Weiss A 1994 *Astron. Astrophys.* **290** 807

[11] Fabian A C and Hansen C J 1979 *Mon. Not. R. Astron. Soc.* **187** 283

[12] Han Z, Podsiadlowski P and Eggleton P P 1995 *Mon. Not. R. Astron. Soc.* **272** 800

[13] Soker N 1998 *Astrophys. J.* **496** 833

[14] Mestel L 1952 *Mon. Not. R. Astron. Soc.* **112** 583

[15] Iben I and MacDonald J 1985 *Astrophys. J.* **296** 540

[16] Harris H C *et al* 2006 *Astron. J.* **131** 571

[17] Spergel D *et al* 2003 *Astrophys. J. Suppl.* **148** 175

[18] Winget D E *et al* 1987 *Astrophys. J. Lett.* **315** L77

[19] Provencal J L *et al* 1998 *Astrophys. J.* **494** 759

[20] Provencal J L *et al* 2002 *Astrophys. J.* **568** 324

[21] Liebert J, Bergeron P and Holberg J B 2005 *Astrophys. J. Suppl.* **156** 47

[22] Lawlor T M and MacDonald J 2006 *Mon. Not. R. Astron. Soc.* **371** 263

# Chapter 21

## Evolution of massive stars

### 21.1 Introduction

Here we consider the evolution of stars that are sufficiently massive such that when helium burning begins, the electrons in the core are non-degenerate. We distinguish between intermediate mass and high mass stars, with the dividing line being whether the star ends its life as a WD or as a neutron star formed in a core-collapse supernova. We have seen earlier that because massive MS stars have convective cores, the evolution across the HRD after H exhaustion in the core is relatively rapid. The core contraction ends when helium burning reactions begin under non-degenerate conditions. The star at this point is a red giant with He burning in a convective core and H burning in a shell outside the He core. Figure 21.1 shows the evolution in the HRD for intermediate mass Pop I stars from the ZAMS to the end of core He burning. Figure 21.2 is the same as figure 21.1 but for high mass stars.

Note the loops to the blue in figure 21.1 that occur in the tracks of the lower mass stars but are absent in the 12 $M_\odot$ track (and also for higher mass tracks). The broken line in figure 21.1 indicates the approximate location of the Cepheid instability strip. $\delta$ Cephei stars show regular oscillations in brightness with periods from 2–40 days. The period is well correlated with luminosity, which makes Cepheid variables very useful as distance indicators. The occurrence of blue loops is very important for the existence of the Cepheid variables. For stars of mass less than 5 $M_\odot$ or greater than 11 $M_\odot$, the tracks cross the instability strip only once in the Hertzsprung gap. For stars of mass between 5 and 10 $M_\odot$, the tracks cross the instability strip three times with the second and third crossings taking much longer than the first crossing. This can be seen in figure 21.3 in which effective temperature is plotted against time for the 6 $M_\odot$ model.

The horizontal broken lines are guides to show that the first crossing is much more rapid than the second and third crossings. Without the blue loops, Cepheid variables would be much less common than observed.

doi:10.1088/978-1-6817-4105-5ch21     21-1

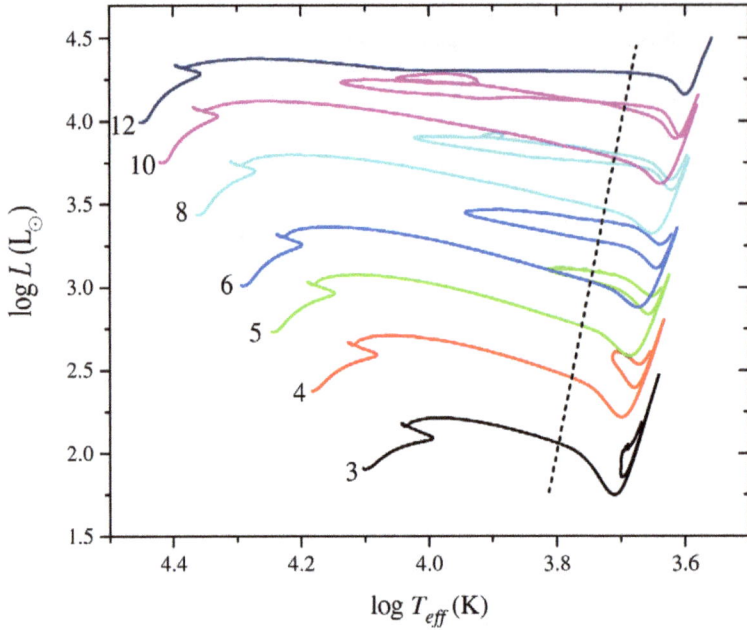

**Figure 21.1.** Evolution in the HRD for intermediate mass Pop I stars from the ZAMS to the end of core He burning. The broken line shows the approximate location of the Cepheid instability strip.

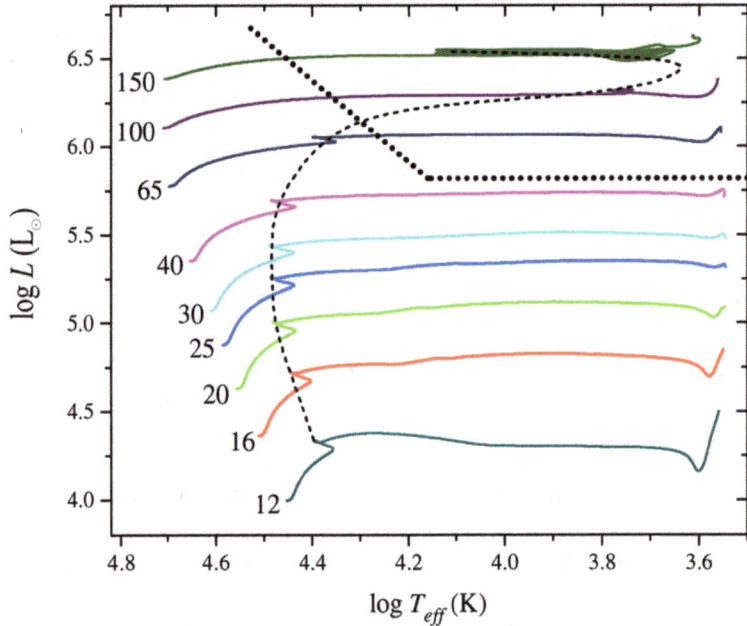

**Figure 21.2.** Evolution in the HRD for high mass Pop I stars from the ZAMS to the end of core He burning. The dashed line indicates where core H burning ends. The dotted line is the location of the Humphreys–Davidson limit.

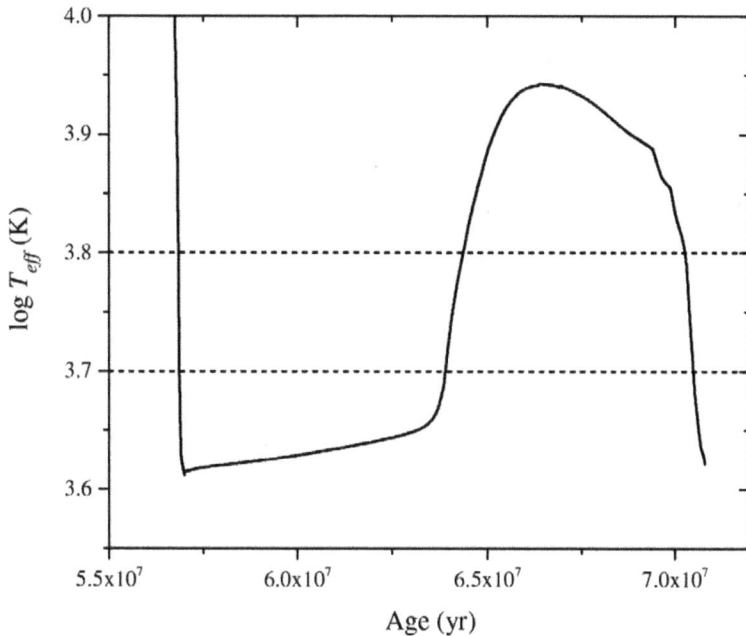

**Figure 21.3.** Cepheid instability strip crossing times.

The dashed line in figure 21.2 indicates where core hydrogen burning ends. The dotted line is the Humphreys–Davidson limit [1]. Humphreys and Davidson found that there is an upper limit to the luminosity of supergiants in the Galaxy and the Large Magellanic Cloud, and that the limit depends on the color of the star. In particular there are no red supergiants with luminosity greater than about $6 \times 10^5 \, L_\odot$. However the evolutionary tracks in figure 21.2 do extend to the red at higher luminosity than this limit. Humphreys and Davidson attributed the lack of very luminous red supergiants to the effects of large amounts of mass loss on the evolution of the most massive stars.

Stars with effective temperatures greater than 30 000 K are of spectral type O. From figure 12.2, we see that they must have an initial mass greater than about $15 \, M_\odot$. Recent studies have shown that most O stars are in binary systems and the stars in these systems are sufficiently close that they will interact at some point in their evolution [2]. This binary interaction may well be responsible for the large amounts of mass loss posited by Humphreys and Davidson to prevent formation of very luminous red supergiants. However about 30% of O stars are single or in binaries with sufficiently large orbital separation that there is no interaction.

Strong evidence that winds remove mass from O stars was found by Morton in 1967 [3]. It was first suggested that these winds result from absorption of photons by ultra-violet resonance lines of highly ionized common elements such as C, N, Si, and S [4]. Later it was shown that the radiation force on resonance lines alone was insufficient to provide the observed wind mass loss rates and that the dominant contribution to the radiative force is due to the large number of subordinate lines of

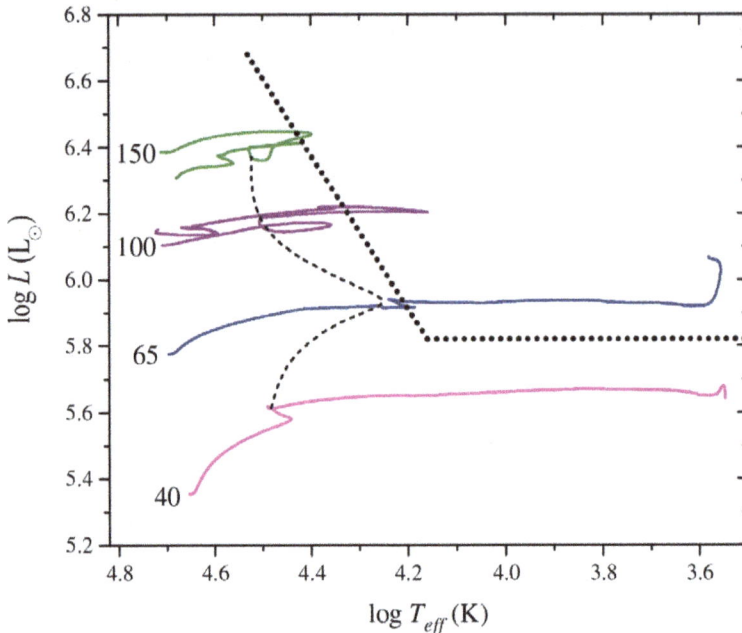

**Figure 21.4.** Evolutionary tracks when wind mass loss is included.

common ions [5]. The theory of radiatively driven winds has developed (see the review [6]) to the point at which theoretical predictions can be compared to measurements of wind mass loss rates from spectroscopic analysis. Once allowance is made for the effects of 'clumping' on the measurements of mass loss rates, good agreement is found with the theoretical predictions of mass loss rates.

We see from figure 21.2 that if winds are responsible for preventing a massive star from becoming a red supergiant, significant mass loss must occur during the MS phase. Figure 21.4 shows some HRD evolutionary tracks for calculations in which wind mass loss is taken into account using theoretically predicted rates for radiatively driven winds [7, 8]. Comparing with figure 21.2, we see that, with this particular mass loss prescription, winds are effective at moving the end of core hydrogen burning to the left of the Humphreys–Davidson limit for masses greater than 65 $M_\odot$. Another consequence of winds is that nuclear processed regions of the star become uncovered due to the mass loss. In massive stars, the CNO-cycles are responsible for converting hydrogen into helium. As we saw earlier, most of the CNO catalysts are converted to $^{14}$N nuclei. Thus once material that once was in the convective cores is uncovered, the surface He and N abundances increase. Once this happens the star becomes a Wolf–Rayet star of spectral type WN, characterized by strong broad emission lines of He and N and, because of high surface temperature, weak lines of H. After the end of core H burning, further mass loss can remove all of the H-rich envelope and the stars becomes a Wolf–Rayet star of spectral type WC or WO, characterized by strong broad emission lines of He, C, and O. Whether this evolutionary sequence from O to WN to WC is a consequence of wind mass loss or

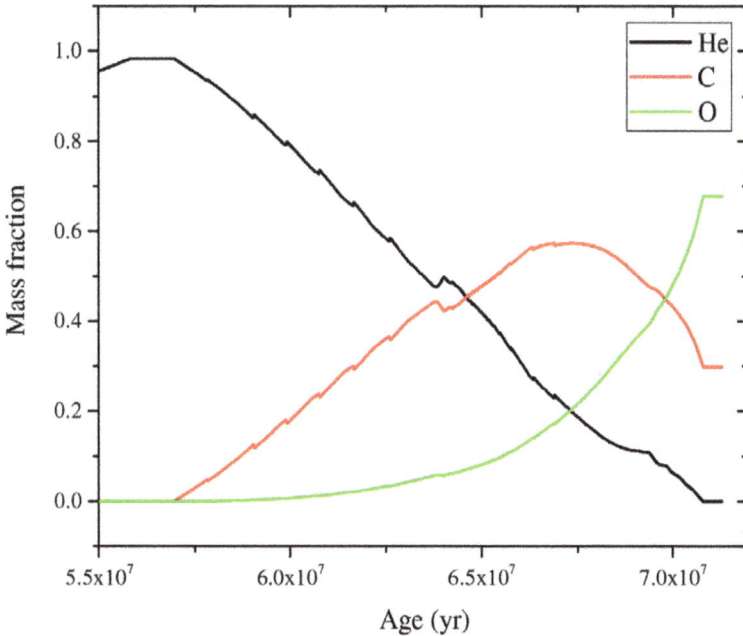

**Figure 21.5.** Composition changes in the core of a 6 $M_\odot$ model during core helium burning.

mass loss due to binary interaction is an active area of research with the current view favoring binary interaction [9].

## 21.2 Composition changes in the core

Initially core He burning proceeds mainly through the $3\alpha$ reaction. As the $^{12}$C abundance increases and the $^4$He abundance diminishes, the $^{12}$C$(\alpha,\gamma)^{16}$O becomes important. Hence the $^{12}$C abundance at first increases in the core, reaches a maximum value, and then decreases again. This is shown in figure 21.5 for a 6 $M_\odot$ model.

## 21.3 Evolution after the end of core helium burning

Because of their convective cores, helium exhaustion in massive stars occurs over an extended region of the core. Hence the core contracts relatively rapidly, until either the electrons become degenerate or carbon burning begins. The critical initial mass that divides these two possibilities is sensitive to the details of convective mixing in the core. In the absence of convective overshooting, this mass is approximately 9 $M_\odot$. For stars in which the electrons in the core become degenerate, the further evolution is similar to that of lower mass stars. The star evolves through a helium shell burning phase followed by a sequence of thermal pulse cycles. As the core grows in mass, the radius and luminosity of the star both increase. This makes it easier for the star to lose mass and the mass loss rate increases with time. The amount of mass lost determines the number of thermal pulses experienced on the TPAGB. The integrated amount of mass loss is constrained by the initial mass–final mass relation for WDs in open clusters.

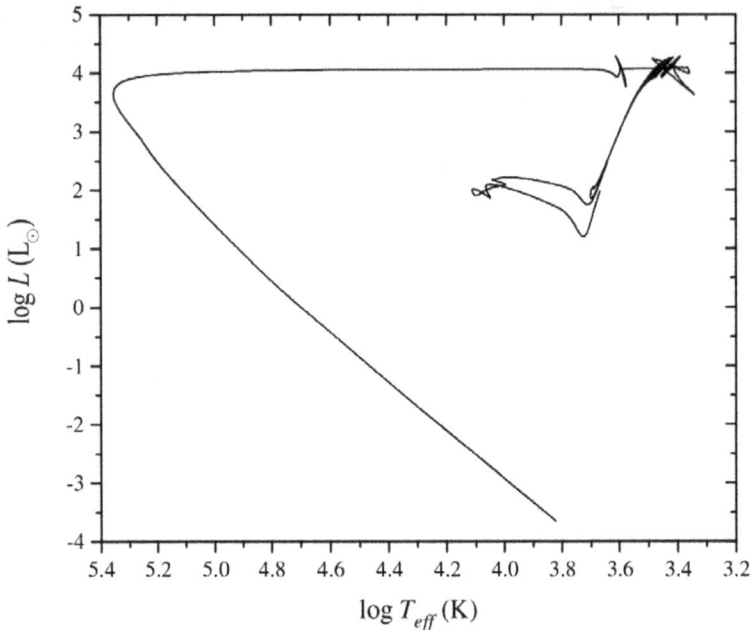

**Figure 21.6.** Complete evolutionary track of a 3 $M_\odot$ Pop I model.

The complete evolutionary track of a 3 $M_\odot$ Pop I model is shown in figure 21.6. An enlargement of the TPAGB phase is shown in figure 21.7.

Figure 21.8 shows the stellar luminosity, the helium burning power and hydrogen burning power against age for the TPAGB phase. We see that the interval between thermal pulses decreases with cycle number and also that the helium burning power increases with cycle number. These are both due to the increasing core mass. The last complete thermal pulse cycle is shown in figure 21.9. We see that the quiescent helium burning phase lasts for about 10% of the complete cycle. This is a consequence of hydrogen fusion producing ten times as much energy per unit mass as helium fusion.

Because of mass loss, the core does not become massive enough for carbon burning to occur and Pop I stars of mass of up to about 8 $M_\odot$ end their lives as WD stars with CO cores [10].

## 21.4 Evolution of stars more massive than 8 $M_\odot$

After the end of core helium burning, the CO core of the star contracts and heats. The density and temperature become large enough that neutrino losses become important. Neutrino losses affect the evolution in different ways depending on the degree of electron degeneracy. If the electrons are non-degenerate, then the energy loss from neutrino processes accelerates the contraction and heating. If the electrons are degenerate the neutrino energy loss results in cooling. Hence if carbon burning does occur, it does so under non-degenerate or only mildly degenerate conditions. For Pop I stars of mass 8 $M_\odot$ and slightly higher, carbon

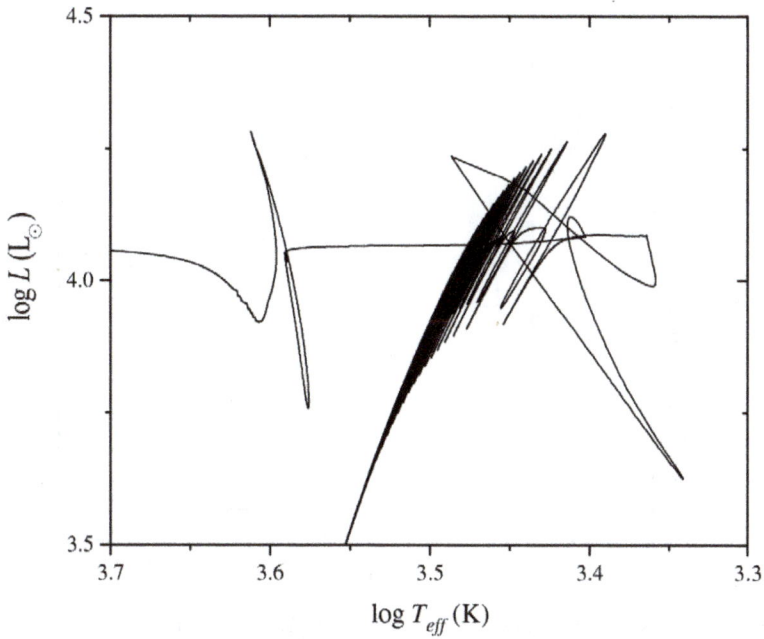

**Figure 21.7.** The TPAGB phase of a 3 $M_\odot$ Pop I model.

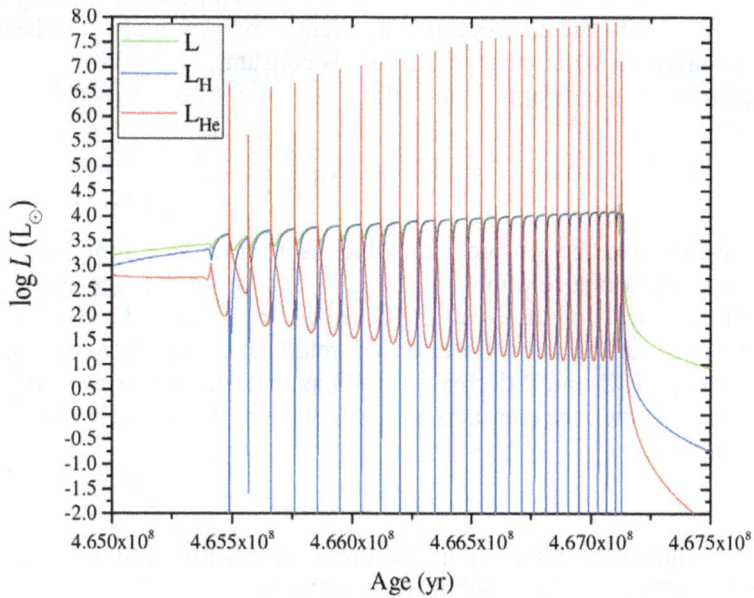

**Figure 21.8.** Stellar luminosity, the helium burning power and hydrogen burning power against age for the TPAGB phase of a 3 $M_\odot$ Pop I model.

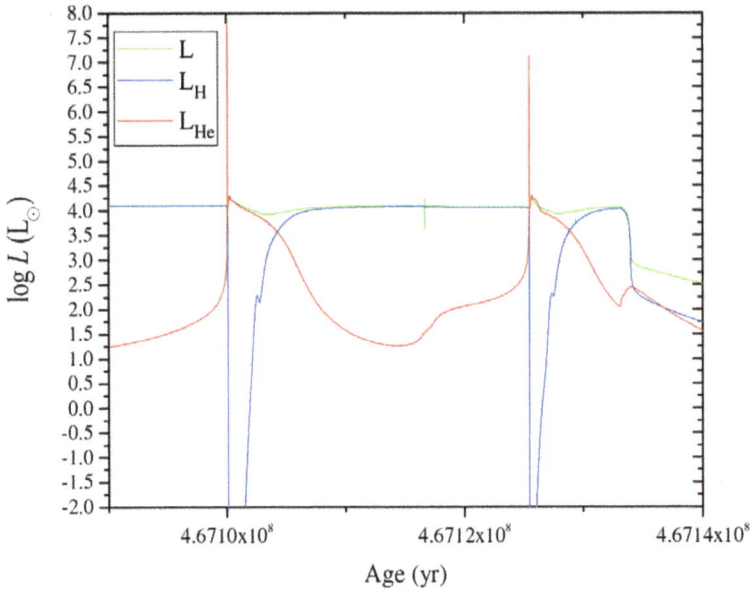

**Figure 21.9.** As figure 21.8, except only the last complete thermal pulse cycle is shown.

ignites off center because the electrons at the very center become degenerate and neutrino losses cool the inner regions.

We can use a simple model to estimate the critical core mass that separates the case in which contraction leads to increasing core temperature from the case in which the electrons in the core become degenerate, preventing further heating. We begin by assuming that the mass of the core is constant.

The equation of state is approximated by

$$P = \Re T \frac{\rho}{\mu_e} + K_\gamma \left( \frac{\rho}{\mu_e} \right)^\gamma. \tag{21.1}$$

Here we have ignored the ion pressure, which is a reasonable approximation since for advanced stages of evolution, $\mu_{ion} \gg \mu_e$. The first term is the non-degenerate electron pressure that dominates at low densities. In the second term, $\gamma$ and $K_\gamma$ are not constants but vary to allow for both non-relativistic and relativistic degeneracy. At low density $\gamma = 5/3$ and it decreases to 4/3 as the electrons become relativistic.

We assume that the core contracts homologously so that the central pressure and density are related by

$$P_c = fGM_c^{2/3} \rho_c^{4/3}, \tag{21.2}$$

where $f$ is a dimensionless constant of order unity. For uniform core density, $f = (\pi/6)^{1/3} = 0.806$, and for a polytrope of index $n$,

$$f = \frac{(4\pi)^{1/3}}{(n+1)(\xi_1^2 \, |\theta'(\xi_1)|)^{2/3}}. \tag{21.3}$$

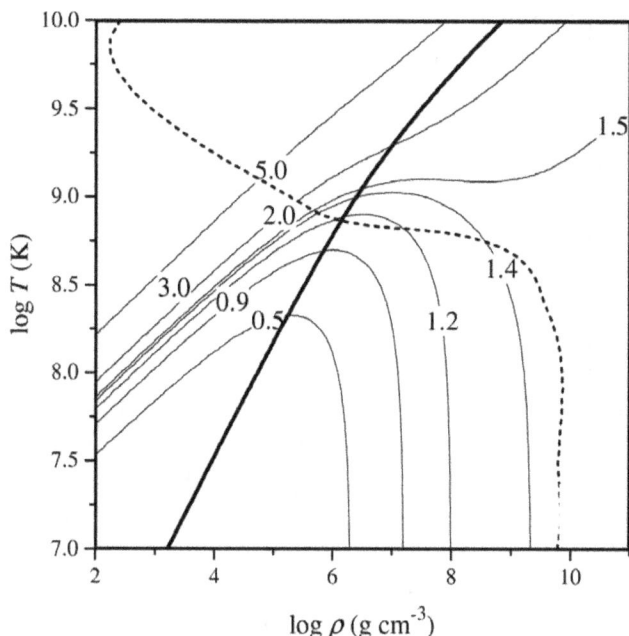

**Figure 21.10.** Simple model for the evolution of a core of fixed mass. The thin solid lines show how the core temperature changes as the core contracts to higher density. The thick solid line shows where the electrons become degenerate and the broken line shows whare carbon begins to burn.

For $n = 1.5$ and $n = 3$, $k = 0.478$ and $0.364$, respectively.

From equations (21.1) and (21.2), we find

$$\Re T_c = f G M_c^{2/3} \mu_e^{4/3} \left(\frac{\rho_c}{\mu_e}\right)^{1/3} - K_\gamma \left(\frac{\rho_c}{\mu_e}\right)^{\gamma-1}. \tag{21.4}$$

If the electrons are non-relativistically degenerate then $\gamma - 1 = 2/3$, and the central temperature has a maximum value, which can be found by setting the derivative of the right-hand side with respect to $\rho_c$ to zero. However if the electrons are relativistically degenerate then $\gamma - 1 = 1/3$, and the central temperature will increase without limit provided the core mass is greater than a critical value which is essentially the Chandrasekhar mass, $M_{ch}$.

The evolution of the core for various masses is shown in the figure 21.10. The thick line is the locus of the transition from non-degenerate to degenerate electrons and the dashed line shows the threshold for carbon burning (taken to be where the nuclear energy generation rate equals the neutrino loss rate). We see that if the core mass, $M_c$, is less than $M_{ch}$, the core temperature reaches a maximum then decreases. This is because the contraction of the core can be halted by electron degeneracy pressure. However if $M_c > M_{ch}$, the temperature increases without limit because electron degeneracy pressure is not sufficient to prevent the contraction. When we consider the nuclear energy production, we see that there is a critical value of $M_c$ (about 1.2 $M_\odot$ in this simple model) below which carbon burning does not start.

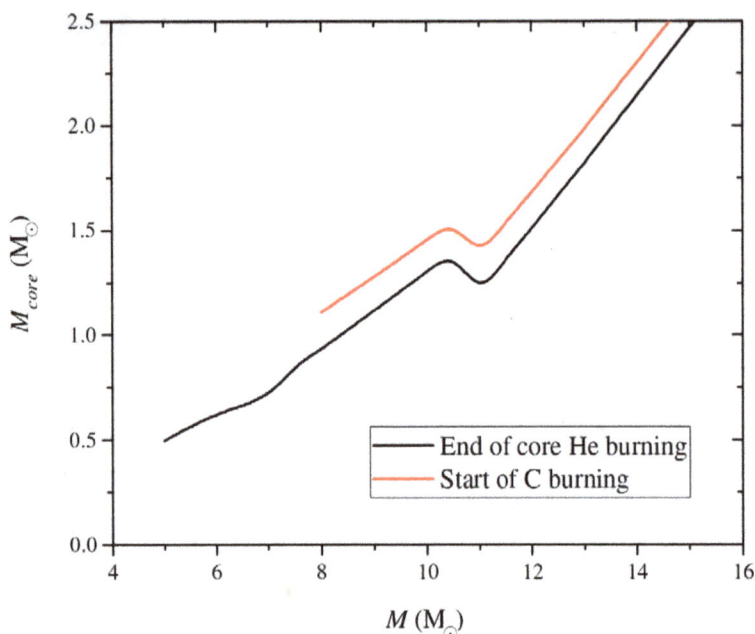

**Figure 21.11.** Core mass at the end of core helium burning and at the beginning of carbon burning in $Z = 0.017$ stellar models.

This critical core mass also places an upper limit on the mass of a CO WD. For larger core masses carbon burning begins under non-degenerate or mildly degenerate conditions.

So far we have considered the case in which the core mass is constant. If the envelope remains sufficiently massive, helium shell burning will increase the mass of the core. Hence if $M_c$ is initially less than about 1.2 $M_\odot$, helium shell burning can increase $M_c$ to the point where carbon burning begins. Because of neutrino losses, the center of the star is cooler than regions further out and so the carbon burning begins off center (this is similar to what happens in core helium flash).

Of course, we cannot directly measure the mass of the core. Figure 21.11 shows the core mass at the end of core helium burning and at the beginning of carbon burning in $Z = 0.017$ stellar models. The dip in core mass for $M \approx 11 M_\odot$ is associated with the absence of blue loops. At lower mass, the blue loops give more time for the helium burning core to grow. For the 8 $M_\odot$ model, the core mass at central helium exhaustion is about 0.93 $M_\odot$ which, from figure 21.10, is less than the critical mass for carbon burning to occur before the electrons at the center become degenerate. The core mass increases due to shell helium burning and carbon burning occurs off center, when the core mass has reached 1.11 $M_\odot$. At lower masses thermal pulses begin before carbon burning can start. Mass loss from winds or envelope ejection (see section 20.4) prevents the core mass from growing to the point at which carbon burning can start. For stars of initial mass between 8 and 10 $M_\odot$, in which carbon burning begins off center, there is a thermally pulsing phase after carbon burning has ended. These stars are called super-AGB stars. Mass loss during the thermally pulsing phase from winds or dynamical instability of the envelope [11] probably removes the outer parts of the star, which ends its life as a CO WD.

In stars of mass greater than about 10 $M_\odot$, carbon burning begins at the center. The subsequent phases of evolution (see section 11.9) are rapid and single stars do not experience significant mass loss during and after carbon burning. If after a particular phase of nuclear burning the core has a mass greater than $M_{Ch}$ the core will contract and heat without the electrons becoming degenerate. The next burning phase will then begin at the center. On the other hand if the core has a mass less than $M_{Ch}$ there is a possibility the electrons in the core become degenerate. In this case shell burning occurs to increase the core mass to the point at which the next fuel ignites off center. After silicon burning the core consists of tightly bound iron peak nuclei, which do not release energy in fusion reactions. The iron core contracts and heats up until the temperature is sufficiently high that energetic photons break up the iron nuclei. Since this process removes energy from the photons and particles responsible for providing pressure support, the core collapses. If the core is not too massive, this collapse is reversed when the density approaches values characteristic of atomic nuclei forming a neutron star. The reversal or 'bounce' leads to a shock wave that propagates through the material outside core (the envelope). Heating from the shock wave leads to nuclear fusion reactions which add energy to the shock wave. Provided the envelope is not too massive, it is ejected in a supernova explosion. The details of the core-collapse supernova mechanism are not yet fully understood, but it seems that formation of the shock wave requires a push from very high energy neutrinos produced in the core collapse interacting with heavy nuclei [12]. Even then this might not be enough for the most energetic core-collapse supernovae [13]. If the envelope is too massive to be ejected, matter will fall back on to the neutron star. Since there is a limit to the mass of a neutron star (this limit is greater than 2.1 $M_\odot$ and probably less than 3 $M_\odot$ [14–16]), further collapse to a black hole can occur. The existence of stellar mass black holes has been inferred from their gravitational attraction in binary systems [17] such as Cygnus X-1. The collapsar model for long duration gamma ray bursts [18] also involves core-collapse.

## Bibliography

[1] Humphreys R M and Davidson K 1979 *Astrophys. J.* **232** 409
[2] Sana H *et al* 2012 *Science* **337** 444
[3] Morton D C 1967 *Astrophys. J.* **150** 535
[4] Lucy L B and Solomon P M 1970 *Astrophys. J.* **159** 879
[5] Castor J I, Abbott D C and Klein R I 1975 *Astrophys. J.* **195** 157
[6] Vink J S 2015 *Astrophys. Space Sci. Libr.* **412** 77
[7] Vink J S, de Koter A and Lamers H J G L 2001 *Astron. Astrophys.* **369** 574
[8] Gräfener G and Hamann W-R 2008 *Astron. Astrophys.* **482** 945
[9] Smith N 2014 *Annu. Rev. Astron. Astrophys.* **52** 487
[10] Doherty C L, Gil-Pons P, Siess L, Lattanzio J C and Lau H H B 2015 *Mon Not. R. Astron. Soc.* **446** 2599
[11] Lau H H B, Gil-Pons P, Doherty C and Lattanzio J 2012 *Astron. Astrophys.* **542** A1

[12] Woosley S and Janka T 2005 *Nat. Phys.* **1** 147
[13] Janka H-T 2012 *Annu. Rev. Nucl. Part. Sci.* **62** 407
[14] Kiziltan B, Kottas A, De Yoreo M and Thorsett S E 2013 *Astrophys. J.* **778** 66
[15] Tolman R C 1939 *Phys. Rev.* **55** 364
[16] Oppenheimer J R and Volkoff G M 1939 *Phys. Rev.* **55** 374
[17] McClintock J E and Remillard R A 2006 Black hole binaries *Compact Stellar X-ray Sources* ed W Lewin and M van der Klis (Cambridge: Cambridge University Press) p 157
[18] MacFadyen A I and Woosley S E 1999 *Astrophys. J.* **524** 262

www.ingramcontent.com/pod-product-compliance
Lightning Source LLC
Chambersburg PA
CBHW081529220326
41598CB00036B/6370